Conserving Farming in the United States

The Methods and Accomplishments of the STEEP Program

Edited by

Edgar L. Michalson
Robert I. Papendick
John E. Carlson

SOIL
AND WATER
CONSERVATION
SOCIETY

CRC Press
Taylor & Francis Group
Boca Raton London New York

CRC Press is an imprint of the
Taylor & Francis Group, an **informa** business

CRC Press
Taylor & Francis Group
6000 Broken Sound Parkway NW, Suite 300
Boca Raton, FL 33487-2742

First issued in paperback 2019

ISBN-13: 978-0-8493-1185-7 (hbk)
ISBN-13: 978-0-367-40007-1 (pbk)

Library of Congress Cataloging-in-Publication Data

Catalog record is available at the Library of Congress

Visit the Taylor & Francis Web site at
http://www.taylorandfrancis.com

and the CRC Press Web site at
http://www.crcpress.com

Preface

This book grew out of the work of the STEEP Program which was begun in 1975 with the first federal funding becoming available in 1976. The Acronym STEEP stands for "Solutions to Economic and Environmental Problems in the Pacific Northwest." This program was organized to develop cooperation between the three Land Grant Universities and the U.S.D.A., Agricultural Research Service research stations in the Pacific Northwest, to establish a new approach to addressing the erosion and water quality problems in the region. Each year erosion losses in the region amount to millions of tons of top soil eroded from its croplands. In some cases, when farmers have used conventional farming practices (heavy tillage) as much as 12 bushels of top soil are eroded annually for each bushel of wheat produced. Average erosion rates in the Palouse region of the Pacific Northwest have ranged from 10 to in excess of 30 tons per acre (approximately $1/8$ inch of topsoil) using conventional practices. Approximately $1/3$ of this eroded soil is washed into streams and other water bodies in the region.

Soil erosion impacts 10 million acres of cropland in the Columbia Plateau, Palouse and Nez Perce prairies, Columbia Basin, and Snake River Plain in Idaho, Oregon, and Washington states. The annual soil erosion in the Columbia River drainage basin at the time the STEEP program was implemented was estimated at 110 million tons. Approximately 30 million tons were estimated to be deposited in Pacific Northwest streams, rivers, lake, reservoirs, and harbors. The resulting build up of silt in reservoirs is projected to shorten the useful life of hydroelectric facilities and irrigation facilities, and had deteriorated the water quality of these bodies. In addition, the cost of removing silt from highway and roadside ditches cost taxpayers millions of dollars annually. In addition to all of these problems, the other consequences of erosion has been that large acreage's of cropland have been denuded of topsoil and their productivity has been seriously reduced.

The causes of causes of soil erosion in the Pacific Northwest have resulted from a combination of factors which include 1) a climate that has a winter precipitation regime, relatively mild winter temperatures, and in some cases a large amount of rain on frozen soil that results in frozen soil runoff; 2) steep and irregular topography of the region; and 3) crop management systems that leave the soil bare going into the winter rainy season. This is especially a problem with fall seeded winter wheat which the major crop in the region.

Since the beginning of the STEEP program, a variety of research and extension programs have been developed which involve cooperation between the three land grant universities and the USDA, along with input from SCS (NRCS), agricultural industry, and farmers in the Pacific Northwest. The funding for this program has come mainly from USDA, CSRS (Cooperative State Research Service of the United

States Department of Agriculture), with some funding coming from the agricultural industry, and commodity commissions in the Pacific Northwest.

One of the major issues which STEEP has addressed is the need to develop long term research which can maintain continuity of effort for a period of time long enough to assure that the results obtained have significant impacts, and that permit the interaction of several disciplines to solve erosion problems. All research projects submitted to be a part of the STEEP program are subject to both per reviews and STEEP reviews. This is done to ensure that only projects of high quality and priority are funded. The effect of this process has been to that researchers have been able to concentrate on addressing erosion problems on a continuous basis with a consistent research output resulting.

The combined research and extension programs of STEEP have developed conservation systems that when implemented do reduce soil erosion significantly. Many of these programs have been adopted by SCS (NRCS) in the implementation of farm conservation plans since the introduction of the Conservation Compliance program in the 1985 farm bill. The number of scientists working on STEEP research tends to vary from year to year depending on the priority needs of the project. The estimates of researchers has varied from about 30 to 70 over the years. The output of the project has been large in terms of the number of publications produced, farmer meetings held, and the annual reviews which are attended by researchers, extension personnel, agri-industry representatives, and farmers.

Under the combined research and extension programs related to STEEP it has been possible to develop conservation systems that are reducing the amount of erosion significantly on Pacific Northwest Farms. Many of the conservation strategies developed have been incorporated into the SCS (NRCS) planning process for conservation compliance which is directed toward reducing erosion in the region to "T" levels. The achievement of "T" levels would reduce erosion to between 3 and 5 tons per acre in the region, compared with the current levels which range between 8 and 30 tons per acre. The research team put together by STEEP has repeated demonstrated it ability to work together to solve erosion problems. The research being addressed deals with implementing no-till farming for small grains and legumes, evaluating new crops and varieties adaptable for conservation tillage, and determining farmers social attitudes concerning the severity of the erosion problems on their farms and in the region. In addition, the development of new conservation systems has involved scientists from 10 disciplines who interact in order to develop consistent research results. STEEP has worked to develop and utilize new and improved systems of conservation management tools in which tillage, crops, plant protection, are all integrated into complete management systems that minimize erosion without adversely affecting production for the farmers.

Finally, the efforts of farmers, extension personnel, SCS (NRCS) conservationists, researchers, and administrators who have worked diligently over the years supporting the STEEP program thanks are due. The STEEP program would never have succeeded without the input of these people. More importantly, the gains achieved by STEEP would never have been made. This book is dedicated to these people and the successes they both individually and cooperatively have achieved over the years of the STEEP program.

Edgar L. Michalson
University of Idaho
Moscow, Idaho

Robert I. Papendick
Washington State University
Pullman, Washington

John E. Carlson
University of Idaho
Moscow, Idaho

Editors

Edgar L. Michalson was professor of Agricultural Economics at the University of Idaho for 27 years until his retirement in 1995. He taught courses in Farm Management, Agribusiness Management, Agricultural Policy, Land Resource Economics, Farm Appraisal, the Economics of Natural Resources, and Research Methodology. His research program included projects on the economics of controlling soil erosion, production economics, natural resource economics, environmental economics, and water resource economics. An important part of his research was devoted to the STEEP program as project leader for the University of Idaho. In this role, he helped faculty develop and coordinate research programs on campus, and worked with the administration in allocating funds to research projects. For 15 years he was co-chairman of the regional STEEP Coordinating Committee with Robert I. Papendick. In this role, he helped coordinate research projects for the regional efforts of STEEP. The major emphasis in this role was to approve new projects and to avoid duplication of effort in researching the causes and factors related to controlling soil erosion.

Robert I. Papendick is formerly Soil Scientist and Research Leader with the USDA Agricultural Research Service at Pullman, Washington. Since coming to Pullman in 1965 his long-term research interest was on no-till of cereals and grain legumes for controlling soil erosion, increasing farm productivity and profitability, and improving soil quality as a way of the future for a more sustainable agriculture. He was one of the original authors of the STEEP program and served as a co-leader of the tri-state effort since its inception in 1975 until 1991. He remained active in the program through research until his retirement from the USDA/ARS in 1995.

John E. Carlson has been employed at the University of Idaho since 1970 and a Rural Sociologist in the College of Agriculture since 1975. His research interests have included wild and scenic rivers, rural health care, the adoption of new technology by farmers, and the social aspects of soil conservation. He established the Social Science Research Unit in the College of Agriculture at the University of Idaho in 1990 and has been its director since that time. He and Don Dillman at Washington State University conducted research on the adoption of soil erosion control practices as part of the STEEP project for about 20 years since the projects beginning in 1975.

Contributors

Alan J. Busacca
Soil Scientist
Crop and Soil Sciences
Washington State University
Pullman, Washington

Gaylon S. Campbell
Soil Scientist
Crop and Soil Sciences
Washington State University
Pullman, Washington

John E. Carlson
Rural Sociologist
University of Idaho
Moscow, Idaho

Peggy M. Chevalier
Crop Physiologist
Crop and Soil Sciences
Washington State University
Pullman, Washington

Hal Collins
Soil Scientist Consultant
Seattle, Washington

Don A. Dillman
Rural Sociologist
Washington State University
Pullman, Washington

Clyde L. Douglas, Jr.
Soil Scientist
USDA-Agricultural Research Scientist
Pendleton, Oregon

Lloyd F. Elliott
Soil Microbiologist (Retired)
USDA-Agricultural Research Service
Corvallis, Oregon

John Hammel
Professor of Soil Science
University of Idaho
Moscow, Idaho

Betty Klepper
Plant Physiologist (Retired)
USDA-Agricultural Research Service
Pendleton, Oregon

Joseph P. McCaffrey
Professor of Entomology
University of Idaho
Moscow, Idaho

Donald K. McCool
Agricultural Engineer
USDA-Agricultural Research Service,
  ARS, PWA
Biological Systems Engineering
Washington State University
Pullman, Washington

Edgar L. Michalson
Agricultural Economist (Retired)
Department of Agricultural Economics
  and Rural Sociology
University of Idaho
Moscow, Idaho

Alex G. Ogg, Jr.
Plant Physiologist (Retired)
USDA-Agricultural Research Service
Pullman, Washington

Robert I. Papendick
Soil Scientist (Retired)
USDA-Agricultural Research Service
Pullman, Washington

Charles L. Peterson
Professor of Biological and Agricultural
  Engineering
University of Idaho
Moscow, Idaho

Keith S. Pike
Entomologist
Washington State University
Prosser, Washington

Sharron S. Quisenberry
Entomologist
University of Nebraska
Lincoln, Nebraska

Paul E. Rasmussen
Soil Scientist
USDA-Agricultural Research Service
Pendleton, Oregon

Richard W. Smiley
Plant Pathologist
Oregon State University
Pendleton, Oregon

Diane E. Stott
Soil Microbiologist
USDA-Agricultural Research Service
National Soil Erosion Research
  Laboratory
West Lafayette, Indiana

Donn C. Thill
Weed Scientist
University of Idaho
Moscow, Idaho

Roger J. Veseth
Extension Conservation Tillage
  Specialist
University of Idaho
Moscow, Idaho

David J. Walker
Professor of Agricultural Economics
University of Idaho

Donald J. Wysocki
Extension Soil Scientist
Oregon State University
Pendleton, Oregon

Douglas L. Young
Professor of Agricultural Economics
Washington State University
Pullman, Washington

Frank Young
Research Agronomist
USDA-Agricultural Research Service
Pullman, Washington

Table of Contents

1 A History of Conservation Research in the Pacific Northwest

Edgar L. Michalson

CONTENTS

1.1 INTRODUCTION

The Inland Empire was created by geologic processes over millions of years. These processes included geological uplift, volcanic activity, erosion, and flooding. Within the geologic time frame, some 10 to 30 million years ago the process of mountain uplifting and volcanic activity began spewing forth lava flows, which eventually covered most of the hills and formed a sea of basalt that is more than 10,000 ft deep in places U.S. Department of Agriculture (USDA, 1978). These basalt flows covered almost all the original valleys and hills. They can still be seen in the river canyons. The next step in the process was dust storms resulting from severe wind erosion in the Columbia Basin area of eastern Washington and Oregon. Loessial soils were deposited to depths of over 200 ft in the Palace area. The wind erosion left the landscape covered with dune-shaped hills with gentle, long south- and west-facing slopes and steep and short north- and east-facing slopes (Pawson et al., 1961).

Among the major geologic effects that formed the region east of the Cascade mountains were the great Spokane Floods (USDA, 1978). These floods were caused by very large ice dams that impounded the Clark Fork River above the present-day location of Missoula, Montana. The melting of the glaciers and great ice fields created very large lakes, and, when the ice dams burst, the flood waters carried away the loess soils. When these dams were breached, three large volumes of water were released, which created the scablands in eastern Washington, and the dry falls on the old Columbia River channel north of Soaplake, Washington. In some areas of eastern Washington the soil was completely removed down to the lava beds. The lands that remained are referred to as the channeled scablands. The lands that were not scoured by these floods became the Palace and the Big Bend area of Washington, the eastern Palace of Idaho, and the eastern Oregon wheat lands.

The rich loessial soils of the region to the east of the Cascades were formed on top of layers of basalt that were generated as a part of the volcanic activity that occurred several millennia ago (USDA, 1978). As this volcanic activity declined, strong winds picked up the fine soils in the Columbia Basin and transported them to what are now called the eastern wheat lands of Washington, Oregon, and northern Idaho. The average depth of the loessial soil mantle in these areas varies from a few inches on the eastern edges of the area to depths of over 100 ft in selected places. This soil-forming process is still going on, albeit at a much reduced scale.

The Palouse soils that were formed were generally well drained, and grasses were the primary form of vegetation in the western, drier areas. In the eastern area, as precipitation increased, shrubs and trees invaded these grasslands. The primary form of vegetation on these lands at the time they were first farmed was Pacific bluebunch wheat grass (*Agropstyron spicatum*). These soils are among some of the most fertile and productive in the world, and it has been boasted by agricultural experts that no other place in the world, under dryland conditions, grows more wheat on a per acre basis (USDA, 1978).

The primary use of these grasslands by the native people was for grazing horses, and as a source of food. Hunting was the second major use. According to V. G. Kaiser, "Prior to the 1870s, the Palouse was: 'Indian Land,' and the use by the white man was confined to a limited amount of fur trapping" (Kaiser, 1961). The native people used the Palace for two major purposes: as an area for hunting upland game animals and birds, which they utilized for food and clothing, and as a grazing area for their rather extensive herds of horses. The native population of the Palouse region was never very large. It has been estimated that the total numbers never exceeded more than 300 at the most (Kaiser, 1978).

The arrival of non-native settlers in the Inland Empire occurred later than did the settlement of western Oregon and Washington. The first non-native people to visit this area were members of the Lewis and Clark expedition. The members of the expedition did not see much of the wheat-growing area because they used the Snake River as their primary means of transport (Durham, 1912). The next white man to come to the Palace was David Thompson, a fur trader and geographer for the British "Northwest Company." The first white man to record crossing the Palouse Hills was John Work who drove a herd of 106 horses from the Lewiston area to Spokane House. His visit was followed by the establishment of fur-trading posts, and later by gold miners, missionaries, and eventually by the railroad surveyors in the 1850s. The man who was probably the first to settle and farm in what is now Whitman County was George Pangburn in 1863 (Babcock, 1940). He settled near the town of Endicott on lower Union Flat Creek, planted a few fruit trees, and started raising livestock.

Serious farming began in the area in the late 1870s and 1880s (Kaiser, 1978). In the higher-rainfall areas, the land was used to produce mainly successive crops of grains. The soil retained its productivity for many years after it was first broken out of sod. Eventually, the productivity was depleted, and grain yields began to decline. During this period the land was worked with horses, and relatively little tillage was done, which minimized the erosion hazard. By 1915, summer fallow became a common practice on fields that had been farmed for 20 years or more (Papendick et al., 1978). At first it was an intermittent practice that allowed the land to lie fallow for a year to rest it. Some of the summer fallow fields were not worked at all. Later on, summer fallowing was used to work the soil to make more plant food available, to control weeds, and to conserve moisture. As summer fallow was introduced into the cropping systems, the erosion hazard became more real. Peas were introduced at the beginning of the World War I period, and where they could be grown they replaced the summer fallow year. To get a suitable seedbed, stubble from the preceding wheat crop was burned in the fall, and the land was plowed. The burning and plowing practice increased the erosion problems greatly compared with the wheat/summer fallow practice. Under the wheat/summer fallow rotation, the wheat stubble was left standing in the field over the winter, which provided some protection for the soil (Kaiser, 1940).

When mechanical power and more-efficient machinery were introduced in the 1920s and 1930s, it permitted a more intensive cultivation of the land (Kaiser, 1961). In 1937, the Washington Agricultural Experiment Station issued Bulletin 344 entitled "Crop Rotations," which in Part I described the effect of rotations on succeeding crops and in Part II described the effects of rotations on the productivity of the soil. Included in Part II was a section on conservation in which it was pointed out that "Losses from soil erosion are so great that both tillage methods and suitable crops should be employed to reduce to a minimum the destruction from this cause" (Wheeting and Vandecaveye, 1937).

In the lower-rainfall areas, summer fallow was a practice used regularly to manage moisture for the grain crops. These lands were settled later on, after the best lands had been taken.

1.2   EROSION PROBLEMS IN EASTERN WASHINGTON AND NORTHERN IDAHO

The history of erosion research in the Pacific Northwest is directly related to national policy on such research. The first serious effort to do something was begun in the Office of Experiment Stations, U.S. Department of Agriculture (USDA) when field studies were begun to find ways of controlling erosion (Wheeting and Vandecaveye, 1937). This was followed in 1914 when field investigations by the Bureau of Agricultural Engineering and the Bureau of Public Roads on terracing were conducted. These studies were continued until 1935. In 1917, the Missouri Agricultural Experiment Station began the first experimental field plots on crop production where runoff and erosion losses were measured. Similar experiments were begun at other locations in the U.S. in 1926 (Horner et al., 1944). The seriousness of erosion in the Inland Empire area of the Pacific Northwest was recognized in the late 1920s, after some 40 to 50 years of cropping, first to continuous wheat, and then to wheat/summer fallow (Kaiser, 1963).

In 1929, Congress appropriated $160,000 for the establishment of ten federal–state erosion experiment stations (Pawson et al., 1961). One of these stations was established at Pullman, Washington in 1930 (USDA, 1978). Following this action, Congress established the Soil Erosion Service (SES) in the Department of Interior in 1933 (Benedict, 1953). In the fall of 1933 the SES established the South Palouse Demonstration Project cooperatively with local farmers. These demonstration projects were located on watersheds near the Palouse Conservation Field Station experimental farm. In 1935, the Soil Erosion Service was transferred to the Department of Agriculture, and all the erosion activities were consolidated into a new agency called the Soil Conservation Service (Benedict, 1953). After this change was made, the North Palouse and Latah-Rock Creek district in Washington and the Latah district in Idaho were voted and organized in 1940, and the Pine Creek district in Washington in 1941 (Pawson et al., 1961). The most important conservation practice begun during this period was the introduction of grasses and legumes into the cropping system to improve the organic matter content of the soils, and to add nitrogen to the soils. Eventually, 40% of the soils were cropped with a grass–legume complex before the introduction of nitrogen fertilizers after World War II.

The war years interfered with the use of conservation practices, particularly with the use of green manure legumes (Kaiser, 1961). Government price supports for peas were very high, and this encouraged farmers to produce this crop. Concern for conservation was ignored to support the war effort. The result of this is that after the war the return to conservation practices was slow at best. Conservation practices were followed in a hit-or-miss fashion and the positive effects of the practices were lost. As a result, most of the benefits of preceding conservation programs were lost.

Research was begun in the early 1930s at the Palouse Conservation Field Station Farm located north and west of Pullman, Washington near Albion, Washington. The first cropping year was 1931–32 and field peas were produced (Horner et al., 1944). The empirical approach to erosion research has measured runoff and erosion soil losses from field plots for over 50 years. The effects of various crop rotations, tillage practices, and management practices were evaluated in terms of the soil loss. Average annual soil losses were not very great from wheat plots in rotations including grasses, legumes, or grass–legume combinations. The most severe erosion consistently occurred on lands that had been summer fallowed before planting winter wheat.

Considerable effort over the years was placed on managing straw residues in the winter wheat summer fallow rotation (Kaiser, 1961). Also, it was observed that soil freezing complicates the effects of tillage practices on runoff and soil losses. It was observed that tillage was often a more important factor than was straw residue management. That is to say, it had a greater effect on erosion and runoff than did straw residues. These data raised the question of the significance of the frozen soil condition on the magnitude of erosion. The problem can be discussed in the following terms, if a soil is wet when it freezes, it becomes very dense, hard, relatively impermeable, and subject to severe erosion. If the same soil is well drained before it freezes, it tends to remain fairly permeable and more resistant to erosion. If the soil is permeable, water tends to be absorbed; if it is not, the water runs off, causing severe erosion. The research efforts on the Palouse Conservation Field Station Farm continue to explore the relationship among tillage, crop residues, and management practices.

Starting in the 1950s, several publications of note were produced cooperatively by USDA-Agricultural Research Service (ARS), Washington State University, and the University of Idaho. Among these was the "Economics of Cropping Systems and Soil Conservation in the Palouse" (Pawson et al., 1961). This study was the first attempt to evaluate the impact of soil erosion economically in the Inland Empire. In addition, other published studies addressed tillage methods for managing straw residues in a winter/wheat summer fallow cropping system (Horner et al., 1944). Terraces were also studied during the 1930s, and it was concluded that they were not practical for the region because of the soils and topography (Wheeting and Vandecaveye, 1937). The results of these studies indicated that runoff and soil losses are reduced by mixing crop residues in the top few inches of the soil, leaving a considerable portion of the residue on the surface of the soil, and by using reduced tillage that leaves the soil surface loose and cloddy which protects it from erosion. All of these practices permit the winter soil moisture to enter the soil profile instead of becoming overland runoff (Wheeting and Vandecaveye, 1937).

Since the 1960s, erosion research at the Palouse Conservation Field Station has concentrated on adapting the "universal soil loss equation" (USLE) to the Pacific Northwest. This effort is continuing with the introduction of the "revised universal soil loss equation" (RUSLE) into the Pacific Northwest region. In addition, the station has studied the effects of soil freezing on erosion. This research is continuing.[1]

1.3   EROSION RESEARCH IN EASTERN OREGON

The Pendelton Experiment Station was organized in the late 1920s. The Oregon Legislature appropriated $2000 annually in 1928 to develop a more profitable system of farming east of the Cascade mountains. On December 1, 1928, a memorandum of understanding between the Oregon Agricultural Experiment Station and the U.S. Bureau of Plant Industry agreed to allocate $9000 of federal money for cooperative research on field crops. In 1928, the Umatilla Court purchased 160 acres at the cost of $30,000 and leased them to the "State Agricultural College of the State of Oregon for use by the Agricultural Experiment Station of the State Agricultural College of the State of Oregon" for special research purposes (Oveson and Besse, 1967).

Over the years since 1928, the Pendelton Station has researched crop rotations to be used in dryland areas; promoted the development of new and improved wheat, barley, and pea varieties; developed new residue-handling systems; used cloddy seedbeds to control soil erosion; developed practical tillage methods, including treatment of land before plowing, plowing, and the cultivation of summer fallow. In addition, the economic use of fertilizers, weed control, seeding rates, new crop evaluation, and the use of weather data have also been studied.

The important work in the period from 1928 to 1966 related to soil erosion control began with the handling of straw residues. This research was done jointly with USDA agricultural engineers who worked on the development of machinery and equipment with adequate vertical and horizontal clearance to avoid clogging, and a shallow and uniform depth of cultivation (Pumphrey and Rasmussen, 1996).

---

[1]"History of Conservation Farming in the Area," source unknown, circa 1950.

Straw spreaders were used to distribute the straw uniformly. If the straw was not uniformly distributed, a stubble buster, skew treaders, or a spike-tooth harrow was used to achieve a uniform distribution of the straw. Research was also conducted on the effectiveness of "stubble mulching in a wheat fallow system" (Oveson and Besse, 1967). Stubble mulching is a year-round way of managing plant residues on cropland. This research demonstrated that sustained high yields can be obtained and soil losses by wind and water erosion can be reduced using stubble mulch practices.[2]

Important research was also done on the use of cloddy seedbeds as a means of controlling erosion. Wheat seeded in clods was tested and it performed as well as that grown in a prepared seedbed (Oveson and Besse, 1967). The use of clods reduced the erosion significantly. Also, the station studied the practice of burning straw before the land was tilled, and the results discouraged the use of this practice. The residue was found to be more valuable in controlling soil erosion than any residual fertilizer value that resulted from the burning. This research also found that the straw residue not only reduced the soil erosion, but it also supplemented the nitrogen supply and improved the tilth of the soil.

After 1967, the emphasis of the Pendelton Experiment Station research program tended to move more strongly in the direction of soil erosion research (Pumphrey and Rasmussen, 1996). However, work did continue in all of the areas discussed above. As chemical controls were developed, weed control and disease control received increased attention. Precipitation, water storage, and erosion research all increased in intensity in the 1950s and 1960s. Wheat straw management, machinery design, tillage, soil fertility, crop rotations, and new crops were studied. Interesting research on the development of the wheat plant was also begun that dealt with the development of computer models that both simulate the growth of the wheat plant and also evaluate management practices to show how they affect the growth and development of the wheat plant.

Soil conservation research has been done at the Sherman Experiment Station located at Moro, Oregon. This station was authorized by the Oregon Legislature in 1909, and research was begun in 1910 (Hall, 1961). Research has been conducted on crop rotations, tillage, fertilizer seeding dates, and the use of tree planting (windrows) to prevent wind erosion. Much of the research on soil erosion at the Sherman Station has been done jointly with the Pendelton Agricultural Research Center.

1.4   EROSION PROBLEMS IN WESTERN OREGON AND WASHINGTON

In the western areas of Oregon and Washington, agricultural production is found mainly in the alluvial river valleys and areas that are subject to severe stream bank erosion. In the cleared upland areas of western Oregon and Washington the soils suffer from sheet, rill, gully, and wind erosion depending upon location, soil type,

---

[2] Unpublished paper, "Wheat Production in the Columbia Basin Counties of Oregon," Pendelton Experiment Station, circa, 1951.

and the nature of farming. The high annual precipitation promotes severe water erosion and excessive leaching of fine soil particles. The upland soils are easily eroded and degraded because they are shallow. Heavy rainfall during the fall and winter and extreme drought during the summer makes farming on lighter soils hazardous. In many areas, large acreages of clean tilled cash crops are grown without any conservation plan in place.

Stream bank erosion is very significant in these areas. It occurs along every river and stream, and is undoubtedly the most important cause of erosion and sediment in the area. The erosion research has been directed largely at watershed management, the use of practices that avoid the washing of soils during the flood seasons, and practices that deal with the heavy precipitation received in this part of the Pacific Northwest region.

1.5 THE HISTORY OF STEEP

In 1975, the directors of the three northwest Agricultural Experiment Stations, the USDA-ARS stations at Pullman, Washington and Pendelton, Oregon, along with the wheat associations and commissions in the region, explored developing a new research effort in the Pacific Northwest. The outcome of their discussions resulted in the development of a tri-state regional soil erosion control project called the STEEP project. STEEP is an acronym for "Solutions to Environmental and Economical Problems," and appended on to this title is, "in the Pacific Northwest." The STEEP project tied together all the above groups into a single research effort. This effort was designed to address the soil erosion problems of the three states. It mobilized the resources of all the major soil erosion interests in the region. In addition, it was developed at a time when the need existed for the region to address this problem. This need is best expressed with the 1972 Amendments to the Water Quality Act. This act had been in existence for some years, and pressure was being applied from the federal level to have the Palouse River meet the standards or the 1982 requirements. It was necessary to demonstrate both that the river was of a quality that the public could use it for contact water sports and that the river would maintain fish life.

The development of the STEEP project was accomplished in the following manner. Farmers in the Pacific Northwest before 1970 had been concerned about what they perceived as declining long-term productivity as reflected by reduced crop yields, particularly on land that had suffered severe erosion. Much of their concern centered on maintaining and, if possible, increasing agricultural efficiency. Each of the three states initiated discussions directed at addressing the broad question of increasing agricultural efficiency. This discussion resulted in several research proposals by the scientists, the wheat growers, and the wheat commissions. When these proposals were reviewed at the national level, it was suggested that a regional research effort be made. In 1974, representatives of the above groups met and began the process of developing the STEEP proposal. The main push to develop the regional effort came largely from the wheat commissions of the states.

The process followed was to develop a process that addressed a large and diverse research problem area. To do this, a large group of scientists were brought together into a regional team effort. Originally, because of the complex biological–climatic–soils system in the Pacific Northwest, it was decided that a 15-year period would be required to carry out enough individual research efforts to address the problem. Generally, these steps were followed:

1. Identify the problem in all its complexity;
2. Determine what needed to be done;
3. Identify individual projects;
4. Determine the resources required and the contribution of each agency;
5. Fund selected projects and avoid duplication of effort;
6. Annually review the progress made.

These steps have been rigorously followed, and continue until the present time to guide the direction of STEEP research.

Two regional committees were set up to provide direction and organization for the STEEP project. The first was a regional Coordinating Committee established at the beginning of the project that has coordinated research efforts. The Coordinating Committee was made up of members from each of the agencies involved. Specifically, the Idaho Agricultural Experiment Station, the Oregon Agricultural Experiment Station, the Washington Agricultural Research Center, the USDA-ARS Pacific Northwest Regional Research Center, the cooperative extensions of the three Northwest states, and USDA-SCS were all represented on the coordinating committee. One of the important functions of the coordinating committee was reviewing all new projects being proposed. The purpose was to avoid duplication of research efforts, and to ensure that new projects would contribute to the goal of controlling soil erosion.

The second committee was the Grower Advisory Committee, which was set up to provide farmers direct input into the STEEP conservation research efforts. This committee meets annually at the STEEP annual review with the coordinating committee and provides important insights and direction for the overall STEEP efforts. Its members are included in the STEEP decision-making process, and they are encouraged to suggest new ideas and directions for the research efforts.

These committees have been highly effective in organizing and providing direction for the STEEP project. The Growers Advisory Committee tells researchers the kinds of problems that farmers face in adopting conservation practices, and what they need in terms of practical approaches to be able to address their conservation problems. The Coordinating Committee also participates in the process and ensures that the grower concerns are included in the research undertaken.

The STEEP project from its inception involved the participation of agricultural researchers from the three Pacific Northwest states, extension personnel, federal researchers, the NRCS (Natural Resource Conservation Service), and farmers. All of these parties joined together to provide direction, insight, and support for the STEEP program. How these groups worked together will be discussed in detail in the following chapter.

The success of the STEEP project is strongly related to the conservation needs of farmers in the area. Since the passage of the 1985 Food and Agriculture Act, which specified that farmers would be required to have conservation plans to receive government payments and other benefits, farmers have renewed their interest in finding economically efficient ways to comply with the conservation provisions of this law. The STEEP project, because it preceded the 1985 act, was already in place and had provided many management practices that farmers adopted to be in compliance. Farmers, NRCS personnel, extension personnel, and environmentalists found the research done by the STEEP program useful in accomplishing the goals of conservation compliance.

The history of conservation research in the Pacific Northwest had its early start during the beginning of this century. Since the 1930s the agricultural colleges and the federal government have been involved in research directed toward minimizing the effects of soil erosion in the region. This effort has been carried on by the STEEP project, and significant progress has been made. Because of the nature of the erosion problem, more research on soil erosion is still needed; although great progress has been made, the erosion problems of the Pacific Northwest have not been solved. Future research on soil erosion will build on the work done by STEEP researchers directed toward the goal of eventually achieving sustainable cropping systems with minimal levels of soil erosion. Society has a real interest in this kind of research, because society is dependent on agriculture for its continuing food supply. In addition, the economic health of the Pacific Northwest is also dependent upon maintaining a stable production system that can sustain itself. Society has a very real stake in the process of encouraging agriculture to develop sustainable production systems to ensure the future of food production in the U.S.

The following chapters will develop in more detail how the STEEP project has worked, will provide examples of the kinds of research that have been done, and will indicate what needs to be done. The authors will discuss management practices and policies that are needed to change existing farming practices and improve farmers' ability to control soil erosion in the Pacific Northwest.

REFERENCES

Babcock, L. T., circa 1940. *Saga of Early History Covering a Portion of Whitman County*, n.p.n.d., p. 133.

Benedict, M. R., 1953. *Farm Policies of the United States, 1790–1950*, The Twentieth Century Fund, New York.

Durham, N. W., 1912. *Spokane and the Inland Empire*, S. J. Publishing Company, Vol. 1, 630.

Hall, W. E., 1961. *Fifty Years of Research at the Sherman Experiment Station*, Agricultural Experiment Station, Oregon State University, Miscellaneous Paper 104, June, Corvallis, Oregon.

Heald, F. D., 1915. *Bunt or Stinking Smut of Wheat*, Agricultural Experiment Station, State College of Washington, Bull. 126.

Horner, G. M., McCall, A. G., and Bell, F. G., 1944. Investigations into Erosion Control and Reclamation of Eroded Land at the Palouse Conservation Experiment Station, Pullman, Washington, 1931–42, USDA Tech. Bull. No. 860.

Horner, G. M., Oveson, M. M., Baker, G. O., and Pawson, W. W., 1961. Effect of Cropping Practices on Yield, Soil Organic Matter, and Erosion in the Pacific Northwest Wheat Region, Agricultural Experiment Stations of Idaho, Oregon, and Washington, and the Agricultural Research Service, USDA, Bull. No. 1.

Horning, T. R. and Oveson, M. M., 1962. Stubble Mulching in the Northwest, USDA-ARS, Soil and Water Conservation Research Division, Agricultural Information Bull. No. 253, March.

Johnson, L. C. and Papendick, R. I., 1968. A brief history of soil erosion research in the United States and in the Palouse, a look at the future, *Northwest Sci.*, 42(2).

Kaiser, V. G., 1940. Report of Agronomic Work of the Soil Conservation Service on the South Fork of the Palouse Demonstration Project: 1934–1939, USDA-SCS, March.

Kaiser, V. G., 1961. Historical land use and erosion in the Palouse—a reappraisal, *Northwest Sci.*, 35(4).

Kaiser, V. G., 1963. Grain recropping, a conservation practice for grainlands in eastern Washington and northern Idaho, *Northwest Sci.*, 37(2).

Kaiser, V. G., circa 1978. A sketch of the agricultural history of the Palouse River Basin, unpublished manuscript.

Oveson, M. M. and Besse, R. S., 1967. The Pendelton Experiment Station—Its Development Program and Accomplishments 1928 to 1966, Oregon Agricultural Experiment Station, Oregon State University, Spec. Rpt. 233, March.

Papendick, R. I., McCool, D. K., and Krass, H. A., circa 1978. Soil Conservation: Pacific Northwest, ASA-CSSA-SSSA.

Pawson, W. W., Brough, O. L., Jr., Swanson, J. P., and Horner, G. L., 1961. Economics of Cropping Systems and Soil Conservation in the Palouse, published cooperatively by the Agricultural Experiment Stations of Idaho, Oregon, and Washington, and the Agricultural Research Service, U.S. Department of Agriculture, Bull. 2, August.

Pumphrey, F. V. and Rasmussen, P. E., 1997. The Pendelton Agricultural Research Center 1967–1992, unpublished manuscript, Agricultural Experiment Station, Oregon State University, 1997.

U.S. Department of Agriculture; Agriculture Research Service, 1979. *Learning from the Past*, October.

U.S. Department of Agriculture, ESCS, FS, and SCS, 1978. *Palouse Co-operative River Basin Study.*

Wheeting, L. C. and Vandecaveye, S. C., 1937. Crop Rotations, Agricultural Experiment Station, State College of Washington, Bull. 344, March.

2 STEEP—A Model for Solving Regional Conservation and Environmental Problems

Robert I. Papendick and Edgar L. Michalson

CONTENTS

Soil erosion continues to be a major problem in the Pacific Northwest wheat-producing region. Each year, erosion losses in the region amount to millions of tons of topsoil eroded from its croplands. A simple calculation shows that with conventional farming as much as 12 bu of topsoil are eroded for each bushel of wheat produced. Average annual erosion rates in the Palouse region of the Pacific Northwest range from 10 to in excess of 30 tons/acre (approximately $1/8$ in of topsoil) using conventional farming practices (USDA et al., 1978). Each year, approximately one third of this eroded soil is washed into streams and other water bodies. This sediment adversely affects many ecosystems and the use of water for industrial and recreational purposes.

Soil erosion impacts 10 million acres of cropland in the Major Land Resource Areas (MLRAs) of the Columbia Plateau, Palouse-Nez Perce Prairies, Columbia Basin, and Snake River Plains in Idaho, Oregon, and Washington. Annual soil erosion in the Columbia River drainage system at the time the STEEP program was implemented in 1975 was estimated at 110 million tons annually (Papendick et al., 1983). Of this amount, approximately 30 million tons annually were deposited in Pacific Northwest streams, rivers, lakes, reservoirs, and harbors (Powell and Michalson, 1986). The resulting buildup of silt in reservoirs was early on projected to shorten the life of hydroelectric and irrigation facilities, and has deteriorated the water quality in these water bodies. In addition, the costs of removing silt from highways and roadside ditches costs taxpayers millions of dollars annually. The other consequence of erosion has been that large acreages of cropland on which the topsoil has been eroded have had their productivity significantly reduced by the loss of topsoil (USDA et al., 1978).

People are becoming more sensitive to the serious consequences of soil erosion, its threat to the environment and the economic security of the region. The capability of farmers to maintain productivity at current levels will depend in large part on their ability to prevent the depletion of organic matter in these soils and future losses of topsoil. The causes of erosion in the Pacific Northwest result from a combination of factors that include (1) a winter precipitation climate with a large amount of frozen soil runoff, (2) the steep and irregular topography of the region, and (3) crop management systems that leave the soil with inadequate surface protection going into the winter rainy season. Management is especially a problem with fall-seeded wheat, particularly when it is planted on summer fallow or harvested dry field pea ground with little surface residue. Although figures are not available, it is a known fact that a high percentage of the annual erosion occurs on land fall-seeded to winter wheat, or is in some way related to winter wheat management practices.

2.1   THE STEEP PROGRAM

STEEP is a coordinated research and education program designed to develop and implement integrated erosion control strategies and practices for the three Pacific Northwest states (Michalson, 1982; Oldenstadt et al., 1982; Miller and Oldenstadt, 1987). The central idea was that soil erosion and water pollution could be significantly reduced by integrating new and improved crop management practices, plant types, pest control methods, and socioeconomic principles into farming systems to achieve sustainable crop production, and a more stable economic environment for farmers. An underlying concept of the STEEP research approach, which has attributed to its success, was finding profitable solutions to environmental problems. The motivation for this research approach came mainly from the wheat producer organizations in the Pacific Northwest. They helped organize the project, provided financial support, obtained supplemental congressional funding, and after inception of STEEP they continued to support and monitor its progress.

Since the start of STEEP, a variety of research and extension programs have been developed which involve cooperation between the three Pacific Northwest Land Grant Universities and the U.S. Department of Agriculture (USDA), Agricultural Research Service (ARS) research facilities at Aberdeen, Idaho, Pendleton, Oregon, and Pullman, Washington. Funds for STEEP research have been made available each year since 1976 by a USDA–Cooperative State Research Service (CSRS) special grant to the agricultural experiment stations (AES) in Idaho, Oregon, and Washington, and by annual appropriations to the USDA-ARS. Researchers in the AES competed for funding under this program on an individual project basis. The approved projects were subject to oversight by a regional coordinating committee, which helped set research priorities and guard against duplication of research efforts. The CSRS and ARS funds were also supplemented by grants from the USDA Soil Conservation Service (SCS, now the Natural Resources Conservation Service, NRCS), agricultural industry, and the commodity commissions in the Pacific Northwest.

One of the issues that STEEP addressed was the need to develop long-term research, which can maintain continuity of the research effort and which permits the interaction of several disciplines to solve erosion problems. The research/education project proposals submitted were all subject to peer review. This was to ensure that the research was targeted at high-priority areas and was of high quality. The overall benefit of the STEEP effort has been that researchers were able to concentrate on addressing erosion problems on a multiyear basis. It ensured a consistent research effort which in the long term paid off in terms of research output. This output consisted of publications dealing with solutions to conceptual problems, professional papers and journal articles, research bulletins, extension publications, software programs, conservation practices, machinery designs, new wheat varieties, and the development of alternative crops for the region. As a part of this process, a consistent effort was made to provide information to the farmers in the region.

Under the combined research and the extension programs within STEEP, it was possible to develop conservation systems that would reduce the amount of erosion significantly on Pacific Northwest farms. Many of the conservation strategies developed have been incorporated into the NRCS planning process for conservation compliance that is directed toward reducing erosion to "T" levels. The achievement of "T" would reduce annual erosion from 12 to over 20 tons/acre on cropland when these conservation practices are applied in the three state region. The research team (approximately 70 scientists) has repeatedly demonstrated its ability to work together to solve erosion problems. STEEP research has dealt with a number of problems including implementing no-till farming for small grains and legumes, evaluating new crops and varieties adaptable for conservation tillage, and determining farmers' social attitudes concerning the severity of the erosion problem on their farms and in their agroclimate zone. The development of new conservation systems has involved university and USDA-ARS scientists from ten disciplines. These scientists have interacted to develop consistent sets of data and models for practices that are compatible with conservation management systems in the region.

The STEEP effort has worked to develop and utilize new and improved systems of conservation management in which tillage methods (crop rotations, plant types, and methods of plant protection) are integrated into complete management systems that minimize erosion without adversely affecting costs and levels of production. There were three main research approaches for erosion control: (1) developing conservation cropping systems along with plant types that can produce profitable yields in trashy hard soil seedbeds; (2) developing early fall planting methods for winter wheat and barley for increased ground cover before winter; and (3) conducting research on the erosion-run process and prediction with emphasis on reducing erosion from frozen and thawing soils. Emphasis is also given to the control of diseases, weeds, insects, and rodents and to socioeconomic factors in the development of these systems.

2.2 STEEP OBJECTIVES

The STEEP research effort was organized into six major objectives (Michalson, 1982; Oldenstadt et al., 1982). Under each objective there are several subobjectives (not listed here). Each subobjective covers a specific research topic that is integrated into the conservation management systems being developed. A brief description of the main objectives follows.

1. *Tillage and plant management.* To develop combinations for tillage and residue management, cropping, and pest control systems to control erosion and maintain and/or increase crop production.
2. *Plant design.* To develop crop cultivars having morphological and rooting characteristics that reduce erosion and that maintain food (feed) production when grown in conservation cropping systems.
3. *Erosion and runoff prediction.* To improve the understanding and prediction of erosion runoff processes as affected by climate, topography, soils, tillage, and crop management; and to use this information in on-farm decision making for planning conservation control and applications.
4. *Pest control.* To integrate the control of weeds, diseases, rodents, and insects into plant management systems to minimize the impacts of changing cropping environments related to erosion control.
5. *Socioeconomics of erosion control.* To estimate the economic and social impacts of using improved erosion control practices on farm organizations, costs and incomes, net income, and on maintaining short- and long-term agricultural productivity in the region.
6. *Integrated technology transfer research.* To develop methods and systems to provide decision support models, software, and published materials to extension agents and specialists, soil conservationists, and producers.

All of the objectives have a common goal, i.e., development of farming systems to control soil erosion. Each objective contributed in a special way to achieve this goal and is a link in the overall project. The objectives and/or subobjectives are revised as necessary as new problems occur in the process of changing tillage and cropping systems.

2.3 ORGANIZATION AND MANAGEMENT OF THE RESEARCH PROGRAM

The research team members repeatedly demonstrated their ability to work together during the tenure of STEEP. The combined effort, tied together by the objectives, coordination among agencies and scientists, and the availability of supplemental funding, demonstrated a high degree of success. A coordinating committee comprised of six scientists, one extension specialist, and a representative from USDA SCS was assigned responsibility for organizing annual reporting sessions. These sessions served as a mechanism for monitoring progress for the wheat industry and participating federal and state agencies, and to facilitate interaction among the scientists. Scientists in the three-state region were encouraged to submit research proposals in their area of interest within the six objectives. These were passed through departmental channels and the research administrators for review and approval. Research grants were usually made for a 3-year period, after which they were terminated or extensively revised. This turnover provided opportunity for different scientists to participate in the program and also a means to maintain a balance in the needed disciplines.

2.4 STEEP EXTENSION

An extension component was added to the STEEP program to disseminate research findings and assist farmers in applying research results. One specialist was located at the Oregon State University Columbia Plateau Conservation Research Center at Pendleton, Oregon and the other at the University of Idaho in Moscow, Idaho. These specialists interacted on a regular basis with scientists, conducted field tours, wrote newsletters, did radio and TV programs, and organized grower information meetings throughout the inland Northwest. One significant quarterly publication by this project was the *STEEP Extension Conservation Farming Update*, authored jointly by the specialists. The STEEP extension component served as a vital link in narrowing the gap between the generation of research information and on-farm application.

STEEP has often been labeled a "growers' program," meaning that individual farmers and the wheat producer organizations provided major input in the operation of the research and extension programs. Not only did the growers seek support for STEEP funding, they also assisted in establishing research priorities through evaluation of existing research projects, and by suggesting new problem areas and needs for future research based on their own experiences. The growers were largely responsible for the addition of the extension component of STEEP.

The wheat grower organizations of the three states did much to obtain the special grant funding that supports STEEP research. By pooling their resources, they convinced Congress of the seriousness of the erosion problem and the need for increased research on erosion control. Their efforts were largely responsible for marshalling the resources of the various research agencies and of the agricultural industry into this high-priority research area.

2.5 STEEP RESEARCH ACCOMPLISHMENTS

The STEEP program contributed a number of major scientific and technical advances in conservation farming since its inception. The following are some examples of these accomplishments.

2.5.1 Tillage And Plant Management

- The yield advantages and improved efficiency of band placement of nitrogen fertilizer was established (Koehler et al., 1987).
- No-till drills were developed with the capability to sow small grains and legumes (lentils, winter peas, etc.) in moderate to heavy crop residues, and in hard dry soils. The drills were also designed to place fertilizer and pesticides (Hyde et al., 1987).
- A crop residue decomposition model using generated residue/soil temperature moisture inputs was developed for predicting the rate at which surface residues disappear in the field (Stroo et al., 1989).
- A wheat growth model was developed that predicts tillage and residue effects on developmental stages of wheat growth (Klepper, 1987).

2.5.2 Plant Design

- Wheat cultivars were screened to determined varietal performance in conservation tillage systems (Allan and Peterson, 1987).
- Risks with early fall seeding for good overwinter ground cover have been reduced by development of wheat types with increased resistance to rusts, foot and root rots, flag smut, and snow mold (Allan and Peterson, 1987).
- Spring wheats that produce only primary tillers were found to be more desirable than secondary tiller–producing varieties in the low-rainfall areas of the Pacific Northwest. In these areas, more spring cropping is encouraged in lieu of summer fallow to control erosion (Konzak et al., 1987).

2.5.3 Erosion And Runoff Prediction

- New factor relationships were developed for the universal soil loss equation (USLE) to improve soil erosion prediction under freeze conditions and steep slopes of the Northwest wheat region. These relationships were used by the SCS in farm planning applications to meet conservation compliance in the 1985 Food Security Act.
- A new method was developed for computing erosivity (R factor) values for rainfall characteristics using hourly rainfall data, which are generally more available than the otherwise required 15-minute "break-point" rainfall data (Istok et al., 1987).
- A rill meter was developed for field measurement of erosion. This tool is the basis of the new data collection procedure needed to develop USLE length of slope (LS) factor relationships for the Pacific Northwest (McCool et al., 1981).

2.5.4 Pest Control

- Conservation tillage practices and intensive cropping to small grains was found to increase root diseases of wheat and barley (Weise et al., 1987).

- Root diseases of wheat and/or barley in a conservation tillage system can be controlled by using a 3-year crop rotation with the cereal grown only 1 year in 3 (Cook and Weller, 1987).
- Improved methods of herbicide management were developed for the control of annual grasses and broadleaf weeds in no-till wheat, barley, and chemical fallow (Rydrych, 1987; Thill et al., 1987).

2.5.5 SOCIOECONOMICS AND PRODUCTIVITY RELATIONSHIPS

- The diffusion of conservation practices in the Northwest will take considerable time and depends to a large extent on changes in the context, innovation, and the characteristics of potential adopters (Dillman et al., 1987).
- STEEP research provided valuable insights for the highly erodible land protection provisions in the 1985 Food Security Act (Hoaf and Young, 1985).
- A farmer survey on adoption of erosion control practices showed that: (1) absentee landowners are not a major constraint to erosion control; (2) the major constraints to erosion control appear to be factors external to the farmer, i.e., especially the rules embedded in government programs; and (3) over the past 20 years there has been a positive change in farmer attitudes toward erosion control as well as implementation of erosion control practices (Dillman and Carlson, 1982; Carlson et al., 1987; 1994).
- STEEP research has shown that wheat yield losses due to loss of topsoil have been masked by technological progress. Improved varieties, fertilizer management, and weed control have all contributed to the masking process by enhancing crop yields (Papendick et al., 1985; Young et al., 1985; Walker and Young, 1986).
- In most cases, net income in the short term was reduced on farms as a result of lower yields with the use of conservation practices (Brooks and Michalson, 1982; Powell and Michalson, 1985; 1986). However, there were options and practices that minimized the short-term losses incurred in the adoption of these practices. Further, the adoption process for conservation practices is a large part of the problem related to the reduced short-run net income. It appears that as farmers gain experience with these practices, crop yields recover after a period of time. Long-term economic studies indicated that economic benefits exist as a result of the soil saved with conservation practices. Reduced tillage and no-till practices on the drier soils of the Pacific Northwest have indicated significant increased moisture storage, and increased crop yields over time. The more erosion is reduced, the greater the long-term productivity of the soil resource (Walker and Young, 1986). More-recent studies have indicated that significant economic benefits are related to the adoption of these practices (Hoaf and Young, 1985). For example, a computer modeling study supported by field data showed that the effect of erosion on wheat yield loss, and the anticipated payoff from future technical progress, is greater on the deep soils than it is on shallower soils (Young, 1984).

2.5.6 Integrated Technology Transfer

- A Pacific Northwest *STEEP Extension Conservation Farming Update* publication is published quarterly by R. Veseth and D. Wysocki.
- A *Pacific Northwest Conservation Tillage Handbook* and a handbook series has also been developed and is available to farmers (Veseth and Wysocki, 1989).
- Software user's manuals have been developed for the GMX program (Bolte and Cornelius, 1989).
- Optional farm plans for conservation compliance have been developed and are being used in the farm planning process (Chvilicek and Ellis, 1988).

2.6   IMPACTS OF STEEP RESEARCH

The bottom-line question is, to what degree have the goals and objectives of the STEEP program been achieved? Has erosion been controlled on Pacific Northwest croplands? Has farm profitability been increased? After 20 years of STEEP, erosion still occurs at unacceptable levels on Pacific Northwest croplands, and some farmers are making less money than they have in the past. Nevertheless, the longer-term benefits of STEEP are becoming evident (Carlson et al., 1994). There has been a visible increase in the adoption of new soil conservation technology that was developed or refined by STEEP researchers. Much of the credit for this is related to the research on fertilizer placement as a part of the no-till drills, which became commercialized in the early 1980s. Conservation methods have enabled farmers to reduce the number of tillage operations on wheat-based rotations from five or more to fewer than three. These methods leave the seedbed in a rough condition and with more surface residues, which reduces erosion.

Much has been learned about the relationships between tillage and plant diseases and how these can be controlled in residue management and crop rotation systems. Methods have been developed to move the pathogen-infested residues away from the seed row. Crop rotations have been designed to reduce adverse biological effects in no-till systems. Economic studies have shown the short- and long-term costs and benefits of soil conservation, and how cost sharing aids in implementation of the best management practices by defraying short-term costs, which otherwise discourage their use by farmers. Overall, STEEP more than any other program has created increased public awareness of soil erosion and its consequences. It has made growers more receptive to implementing conservation measures on their farms.

2.7   FUTURE DIRECTIONS FOR STEEP

In spite of the technological advances made by STEEP, soil erosion is still a major environmental and economic problem in the region. Developing better conservation practices and achieving more widespread application on the land is still needed. The mandatory conservation compliance requirements of the 1985 Food Security Act and policies of subsequent Farm Bills will likely speed up adaptation of conservation tillage practices such as no-till, and many growers have been testing these practices during the implementation phase of the Food Security Act. Many technical problems relating to the use of these practices have not yet been solved, and consequently their final economic and social acceptability are not adequately known. For example, many farmers have had difficulty planting and harvesting crops, and controlling weeds, diseases, and other pests in seedbeds that are rough tilled or contain large amounts of surface residues. These obstacles can frequently be overcome in research plots, but not always on farmers' fields for many reasons including lack of know-how, inadequate equipment, cost or time limitations, or variations in soil character-istics. Much remains to be done to accomplish the adoption of conservation tillage technology on the farms in the region.

2.8   EMERGING ISSUES

Concerns in addition to soil erosion have now placed land stewardship in a new context. Water and air quality have become major national issues. In the future, increasingly severe restrictions on chemical and nutrient management and control of wind erosion are likely to be legislatively mandated. Fewer agricultural chemicals will be available to farmers, particularly for minor-use crops because of increased registration restric-tions and expenses. Other issues that farmers must face include energy conservation, environmental stability, fish and wildlife protection, farmworker health, food safety, and maintaining net farm incomes. Conservation farming for erosion control affects all of these issues and concerns. For example, with current technology, conservation practices such as reduced tillage and no-till may require increased use of pesticides and nitrogen fertilizer. Reduced runoff and evaporation under conservation tillage may increase infiltration and leaching of chemicals into groundwater. If this happens, some of the chemical-intensive practices in use today will not be acceptable in the future. Thus, the concerns for efficient crop production, conservation of soil, water, and energy, and maintenance of net farm income will have to be integrated with the need to safeguard human health and protect the environment.

About two million acres of highly erodible land in the three Pacific Northwest states is now in the Conservation Reserve Program (CRP). This land is under 10-year contracts to grow only with grass or trees, which do not require tillage or harvest. Much of this land had been seriously degraded because of excessive erosion rates with the cropping systems previously used. If these lands are converted back to crops after the CRP contracts expire, farmers will be faced with the challenge of conserving and maintaining the accrued productivity benefits. This will require the use of cropping systems that are more sustainable than those used prior to placing these lands in the reserve. The outcome of this post-CRP transition will have a profound impact on soil erosion, water quality, farm income, and the future of agriculture in the region.

2.9   CONCLUDING REMARKS

Nationally, the momentum for conservation is accelerating. The STEEP program for the Pacific Northwest continues to be the best proven vehicle for coordinating the research and extension efforts needed to assure that national and regional goals for resource protection and economic viability are achieved. STEEP as a research/extension program has a proven track record related to organizing and accomplishing research goals in the Pacific Northwest. STEEP II and now STEEP III, its follow-on, build on this foundation to help farmers meet the requirements of conservation compliance, and maintain profitable operations as agriculture moves into the next century.

REFERENCES

Allan, R.E. and Peterson, C.J., Jr., 1987. Winter wheat plant design to facilitate control of soil erosion, in L.F. Elliott, R.J. Cook, M. Molnau, R.E. Witters, and D.L. Young, Eds., *STEEP—Conservation Concepts and Accomplishments*, Pub. 662, Washington State University, Pullman, 225–245.

Bolte, J.P., Lev, L., and Cornelius, J., 1989. *GMX Users Manual*, Department of Agricultural Engineering, Corvallis, OR.

Brooks, R.O. and Michalson, E.L., 1982. An Evaluation of Best Management Practices in the Cow Creek Watershed, Latah County, Idaho, Idaho Agricultural Experiment Station, Research Bulletin No. 127, University of Idaho, Moscow.

Carlson, J.E., Dillman, D.A., and Boersma, L., 1987. Attitudes and behavior about soil conservation in the Pacific Northwest, in L.F. Elliott, R.J. Cook, M. Molnau, R.E. Witters, and D.L. Young, Eds., *STEEP—Conservation Concepts and Accomplishments*, Pub. 662. Washington State University, Pullman, 333–341.

Carlson, J., Schnabel, B., Bews, C.E., and Dillman, D.A., 1994. Changes in the soil conservation attitudes and behaviors of farmers in the Palouse and Camas Prairies: 1976–1990, *J. Soil Water Conserv.*, 49, 493–500.

Chvilicek, J. and Ellis, J.R., 1988. Feasibility of conservation compliance in the Palouse region of the Pacific Northwest, paper presented at the Western Agricultural Economics Association Annual Meeting, Honolulu.

Cook, R.J. and Weller, D.M., 1987. Management of take-all in consecutive crops of wheat and barley, in I. Chet, Ed., *Nonconventional Methods of Disease Control*, John Wiley & Sons, New York, 41–76.

Dillman, D.A. and Carlson, J.E., 1982. Influence of absentee landlords on soil erosion control practices, *J. Soil Water Conserv.*, 37, 37–40.

Dillman, D.A., Beck, D.M., and Carlson, J.E., 1987. Factors affecting the diffusion of no-till agriculture in the Pacific Northwest, in L.F. Elliott, R.J. Cook, M. Molnau, R.E. Witters, and D.L. Young, Eds., *STEEP—Conservation Concepts and Accomplishments*, Pub. 662, Washington State University, Pullman, 343–364.

Hoaf, D. and Young, D., 1985. Toward effective land retirement legislation, *J. Soil Water Conserv.*, 40, 462–465.

Hyde, G.M., Wilkins, D.E., Saxton, K.E., Hammel, J.E., Swanson, G., Hermanson, R.E., Dowding, E.A., Simpson, J.B., and Peterson, C.L., 1987. Reduced tillage seeding equipment, in L.F. Elliott, R.J. Cook, M. Molnau, R.E. Witters, and D.L. Young, Eds., *STEEP—Conservation Concepts and Accomplishments*, Pub. 662, Washington State University, Pullman, 41–56.

Istok, J.D., Zuzel, J.F., Boersma, L., McCool, D.K., and Molnau, M., 1987. Advances in our ability to predict rates of runoff and erosion using historical climatic data, in L.F. Elliott, R.J. Cook, M. Molnau, R.E. Witters, and D.L. Young, Eds., *STEEP—Conservation Concepts and Accomplishments*, Pub. 662, Washington State University, Pullman, 205–222.

Klepper, E.L., 1987. Crop response to tillage, in R. Papendick, Ed., *Tillage for Crop Production in Low Rainfall Areas*, Agricultural Services Division of the Food and Agricultural Organization of the United Nations, Rome, Italy.

Koehler, F.E., Cochran, V.L., and Rasmussen, P.E., 1987. Fertilizer placement nutrient flow, and crop response in conservation tillage, in L.F. Elliott, R.J. Cook, M. Molnau, R.E. Witters, and D.L. Young, Eds., *STEEP—Conservation Concepts and Accomplishments*, Pub. 662, Washington State University, Pullman, 57–65.

Konzak, C.F., Sunderman, D.W., Polle, E.A., and McCuiston, W.L., 1987. Spring wheat plant design for conservation tillage management systems, in L.F. Elliott, R.J. Cook, M. Molnau, R.E. Witters, and D.L. Young, Eds., *STEEP—Conservation Concepts and Accomplishments*, Pub. 662, Washington State University, Pullman, 247–273.

McCool, D.K., Dossett, M.G., and Yecha, S.J., 1981. A portable rill meter for field measurement of soil loss, in *Erosion and Sediment Transport Measurement, Proceedings of the Florence Symposium*, June, IAHS Publ. No. 133.

McCool, D.K., Zuzel, J.F., Istok, J.D., Formanek, G.E., Molnau, M., Saxton, K.E., and Elliott, L.F., 1987. Erosion processes and prediction for the Pacific Northwest, in L.F. Elliott, R.J. Cook, M. Molnau, R.E. Witters, and D.L. Young, Eds., *STEEP—Conservation Concepts and Accomplishments*, Pub. 662, Washington State University, Pullman, 187–204.

Michalson, E.L., 1982. STEEP, A Multidisciplinary Multi-Organizational Approach to Soil Erosion Research, Department of Agricultural Economics, AE Research Series 238, University of Idaho, Moscow.

Miller, R.J. and Oldenstadt, D., 1987. STEEP history and objectives, in L.F. Elliott, R.J. Cook, M. Molnau, R.E. Witters, and D.L. Young, Eds., *STEEP—Conservation Concepts and Accomplishments*, Pub. 662, Washington State University, Pullman, 1–7.

Oldenstadt, D.L., Allan, R.E., Bruehl, G.W., Dillman, D.A., Michalson, E.L., Papendick, R.I., and Rydrych, D.J., 1982. Solutions to Environmental and Economic Problems (STEEP), *Science*, 217, 904–909.

Papendick, R.I., McCool, D.K., and Krauss, H.A., 1983. Soil conservation: Pacific Northwest, in H.E. Dregne and W.O. Willis, Eds., *Dryland Agriculture*, Agonomy Monograph 23, American Society of Agronomy, Crop Science Society of America, and Soil Science Society of America, Madison, WI, 273–290.

Papendick, R.I., Young, D.L., McCool, D.K., and Krauss, H.A., 1985. Regional effects of soil erosion on crop productivity—the Palouse area of the Pacific Northwest, in R.F. Follett and B.A. Stewart, Eds., *Soil Erosion and Crop Productivity*, American Society of Agronomy, Crop Science Society of America, and Soil Science Society of America, Madison, WI, 305–320.

Powell, M.L. and Michalson, E.L., 1985. An Evaluation of Best Management Practices on Dryland Farms in the Lower Portion of the Snake River-Basin of Southeastern Idaho, Idaho Agricultural Experiment Station, Bulletin 639, University of Idaho, Moscow.

Powell, M.L. and Michalson, E.L., 1986. An Evaluation of Best Management Practices on Dryland Farms in the Upper Snake River Basin of Southeastern Idaho, Idaho Agricultural Experiment Station, Bulletin 655, University of Idaho, Moscow.

Rydrych, D.J., 1987. Weed management in wheat—fallow conservation tillage systems, in L.F. Elliott, R.J. Cook, M. Molnau, R.E. Witters, and D.L. Young, Eds., *STEEP—Conservation Concepts and Accomplishments*, Pub. 662, Washington State University, Pullman, 289–298.

Stroo, H.F., Bristow, K.L., Elliott, L.F., Papendick, R.I., and Campbell, G.S., 1989. Predicting rates of wheat residue decomposition, *Soil Sci. Soc. Am. J.*, 53, 91–99.

Thill, D.C., Cochran, V.L., Young, F.L., and Ogg, A.G., Jr., 1987. Weed management in annual cropping limited-tillage systems, in L.F. Elliott, R.J. Cook, M. Molnau, R.E. Witters, and D.L. Young, Eds., *STEEP—Conservation Concepts and Accomplishments*, Pub. 662, Washington State University, Pullman, 275–287.

U.S. Department of Agriculture, Soil Conservation Service, 1972. Major Problem Areas for Soil and Water Conservation in the Dry Farmed Grain Lands of the Columbia-Snake-Palouse Area of the Pacific Northwest, A Field Report, Soil Conservation Service, Spokane, WA.

U.S. Department of Agriculture, Economics, Statistics and Cooperative Service, Forest Service, and Soil Conservation Service, 1978. *Palouse Cooperative River Basin Study*, U.S. Government Printing Office, Washington, D.C., 658–797.

Veseth, R.J. and Wysocki, D.J., 1989. *Pacific Northwest Conservation Tillage Handbook*, PNW Misc. Pub. 0136, Washington State University, Pullman, University of Idaho, Moscow, and Oregon State University, Corvallis.

Walker, D.J. and Young, D.L., 1986. Effect of technical progress on erosion damage and economic incentives for soil conservation, *Land Econ.*, 62, 89–93.

Weise, M.V., Cook, R.J., Weller, D.M., and Murray, T.D., 1987. Life cycles and incidence of soilborne plant pathogens in conservation tillage systems, in L.F. Elliott, R.J. Cook, M. Molnau, R.E. Witters, and D.L. Young, Eds., *STEEP—Conservation Concepts and Accomplishments*, Pub. 662, Washington State University, Pullman, 299–313.

Young, D.L., 1984. Modeling agricultural productivity impacts of soil erosion and future technology, in B.C. English, J.A. Maetzold, B.R. Holding, and E.O. Heady, Eds., *Future Agricultural Technology and Resource Conservation*, Iowa State University Press, Ames, 60–85.

Young, D.L., Taylor, D.B., and Papendick, R.I., 1985. Separating erosion and technology impacts on winter wheat yields in the Palouse: a statistical approach, in *Erosion and Soil Productivity, Proceedings of the National Symposium on Erosion and Soil Productivity*, December, 1984, New Orleans, American Society of Agricultural Engineers, St. Joseph, MO, 130–142.

3 Measuring and Modeling Soil Erosion and Erosion Damages

Donald K. McCool and Alan J. Busacca

CONTENTS

3.1 INTRODUCTION

3.1.1 BACKGROUND

The majority of soils of the Northwestern Wheat and Range Region (NWRR) (Austin, 1981) are formed from loess deposits. Soils derived from loess extend over a vast area on the Columbia Plateau in eastern Washington, northern Idaho, and northcentral Oregon and on the Snake River plain in southern Idaho. The Palouse is the geological center of the loess deposit on the Columbia Plateau; it reaches a thickness of more than 250 ft. (75 m) in Whitman County, Washington. In other parts of the NWRR, such as in northcentral Oregon, in northern Idaho, and in

Lincoln, Adams, and Douglas Counties in Washington and the Snake River plain, the loess ranges from a few feet to 50 ft. thick. The loess has accumulated from seasonal dust storms over about the last 2 million years (Busacca, 1989; 1991). The upper 2 to 5 ft. of loess that forms the modern surface soil has accumulated since about 18,000 years ago (Busacca and McDonald, 1994; Berger and Busacca, 1995; Richardson et al., 1997).

The topography of the NWRR is distinctive; the slopes range from gentle to more than 50%, and the hills have a variety of complex shapes and patterns. In fact, the complexity in detail of landscapes led Kaiser et al. (1961) to subdivide the Palouse into 12 different landform types that have somewhat different land uses. The most complex shapes and patterns are found in the main Palouse because the great height of the loess overlying the basalt bedrock increases the impact of water runoff. The unusual landforms are the result of an interplay between deposition of incoming dust, which builds or enlarges the hills, and geologic erosion by water and wind, which wears them down (Caldwell, 1961; Busacca et al., 1985; Busacca, 1991).

Specific variations in the topography result from several factors: the texture of the loess (proportions of sand, silt, and clay), wind intensity and direction, amount and intensity of precipitation (rain and snow), amounts of seasonal snow accumulation, frequency and depth of soil freezing, and total loess thickness above basalt bedrock (Busacca et al., 1985). For example, the drier western Palouse is characterized by long linear hills formed by processes of wind deposition and wind erosion in moderately thick sandy loess, whereas the wetter eastern Palouse is characterized by steep, complex hill shapes (Mulla, 1986) formed by wind deposition of finer loess modified by water erosion, snow melt on frozen soil (Kirkham et al., 1931; Rockie, 1950), and mass movements such as mudflows and landslides (Kardos et al., 1944).

The NWRR has a Mediterranean-type seasonal rainfall pattern with cold wet winters and warm to hot, dry summers. A strong climatic gradient exists: mean annual precipitation (MAP) increases from about 9 in. (230 mm) in the wheat-fallowed southwestern part to more than 25 in. (635 mm) in the annual-cropped northeastern part. The percentage of the mean annual precipitation falling as snow increases from southwest to northeast. Mean annual temperature (MAT) and length of the growing season both decrease across this transect.

Natural vegetation in the region at the time of the first non-native-American settlers was a steppe or prairie consisting dominantly of perennial bunchgrasses. The drier western part supported sagebrush–fescue or sagebrush–wheatgrass (*Artemisia–Festuca* or *Artemisia–Agropyron*) plant communities, whereas the wetter eastern parts supported wheatgrass–fescue or fescue–snowberry (*Agropyron–Festuca* or *Festuca–Symphoricarpos*) plant communities (Daubenmire, 1970). Conifers created an open "parkland" forest in a mosaic with grasses and shrubs east of Moscow, Idaho, and dense conifer stands covered the easternmost extension of the loess in northern Idaho. Most of these timbered areas have been cleared for farming.

3.1.2 The Finite Soil Resource

Perhaps the greatest misperception about the soil resource of the NWRR is that it is inexhaustible. This error is understandable given the tremendous thickness, up to

250 ft. (Ringe, 1970), and the vast extent of the loess. Soil productivity is determined, however, by the properties of the topsoil, and the rooting zone for plants, about 5 ft. or less. The beneficial properties of the topsoil (which include high organic matter content, water- and wind-stable soil aggregates, high porosity and permeability for root and water penetration, and high water-holding capacity) are progressively being lost under annual cultivation (Pikul and Allmaras, 1986; Rasmussen and Parton, 1994) and accelerated erosion. Subsoils restrictive to root and water penetration are common in the NWRR, and accelerated erosion brings them to the surface.

The surface soils have formed in the newest layer of loess that has been deposited during the last 18,000 years. This "skin" of young loess is responsible for the natural soil productivity. The topsoil (A horizon or *mollic epipedon*) of uneroded NWRR soils is dark brown to black silt loam that has granular structure and is high in humified organic matter because it is the principal zone of plant rooting, nutrient cycling, and biological mixing. Organic carbon content and thickness of A horizons before cultivation ranged from about 0.5% and 8 to 12 in., respectively, in the driest areas of loessial soils to 5.0% and 20 to 36 in. in the highest-rainfall former grassland parts of the eastern Palouse (Lenfesty, 1967; Donaldson, 1980; Gentry, 1984; Rodman, 1988). Most soils have a *cambic* subsoil, or B horizon, that is yellowish brown silt loam with blocky structure. Two examples of deep fertile soils found where the young loess is 5 ft. or more thick are the Walla Walla and Palouse series. Soils such as these generally are the most productive and simplest to manage because the effective rooting depth, fertility, and water storage capacity remain very high even if they have been eroded.

Soil depth and soil properties differ from one part of the landscape to another even in unfarmed areas because the thickness of the youngest loess layer varies across hill slopes and across the region, as does the character of older loess beneath the youngest loess. In places where the young loess was naturally very thin, such as on hilltops and convex midslope knobs, or where accelerated erosion has been severe (or both), the topsoil and cambic horizons are thin or absent and older subsoil materials are near or at the surface. The older loess beneath the surface soil generally is much less favorable for plant growth. This is because the older loess contains ancient buried soils called paleosols (*paleo* = ancient; *sol* = soil). Most paleosols are more dense, less permeable, and less fertile than are the surface soil horizons due to lime and silica cementation (*duripans*) in drier areas or clayey textures (*argillic horizons* and *fragipans*) in higher-rainfall areas. When paleosol features or horizons occur within 6 ft. or less of the surface, they are important because they begin to reduce rooting depth and plant-available water. Examples of soils that have restrictive paleosol horizons in their lower part are the Endicott, Thatuna, and Santa series in dry, intermediate, and moist zones, respectively. More than one half of the soil series in the soil survey of Whitman County, Washington (Donaldson, 1980), and about 30% of the upland acres there have restrictive paleosol horizons in the rooting zone (Busacca et al., 1985).

In some parts of both the native and farmed landscapes, the paleosol horizons or fragments of the paleosol materials may be at or only inches beneath the surface. This is the case on ridge tops and south-facing slopes where both geologic erosion and accelerated water erosion plus tillage soil movement are severe (USDA, 1978;

Busacca et al., 1985; Rodman, 1988; Montgomery, 1993; McCool et al., 1997). Examples are soils such as the Lance series in the low-rainfall zone and Naff and Garfield series in the intermediate-rainfall zone. The U.S. Department of Agriculture (USDA) estimated in 1978 that accelerated water erosion over the last 100 years has removed all of the original topsoil from 10% of Palouse cropland and has removed $^{1}/_{4}$ to $^{3}/_{4}$ of the topsoil from another 60% of the cropland (USDA, 1978). Despite adoption of soil-conserving techniques since 1978, such as divided slope farming, minimum tillage systems, and increased crop residue cover, continued soil loss by erosion and tillage are diminishing topsoil depth and increasing the acreage of exposed paleosol subsoils, putting the long-term sustainability of the soil resource base in question.

3.1.3 MAGNITUDE AND NATURE OF THE EROSION PROBLEM

The nonirrigated cropland of the NWRR has long been recognized as an area with severe erosion problems. Much of the 10 million acres of nonirrigated cropland is subject to wind or water erosion, or both. There is great variation in water erosion rates in the region due to climatic, topographic, soil, and management differences. In contrast to the Great Plains and other regions of the U.S. east of the Rocky Mountains where a summer rainfall prevails, winter climate dominates the water erosion process. Much of the water erosion is caused by rain or snowmelt on thawing soils which, for much of the area, can occur from December through March, rather than at the end of the winter season as in areas with colder climates. Whitman County, Washington, in the heart of the region, has historically high rates of soil erosion. Annual erosion by water on the 1.1 million acres of cropland in Whitman County in the 43 years from the winters of 1939–40 through 1981–82 was estimated to range from 0.5 to 21.9 million tons (personal records of Verle G. Kaiser). The mean annual soil erosion was 9.6 million tons, for a mean annual soil erosion of about 24 ton/acre of seeded winter cereal. These rates include sheet, rill, and concentrated flow erosion.

Wind erosion affects lower precipitation irrigated and nonirrigated cropland of the NWRR. The soils in this area are coarse textured and can suffer severe erosion in the fall and early spring unless protected by plant cover or crop residue. About 6.8 million acres are susceptible. The eroded material damages growing crops, reduces soil productivity, and creates health, visibility, and other air quality problems, particularly in more heavily populated areas such as Spokane, Coeur d'Alene, Richland, Kennewick, and Pasco (Papendick and Veseth, 1996). Wind erosion prediction technology is undergoing rapid changes as a result of recent emphasis on solving these problems (Saxton et al., 1996; Stetler and Saxton, 1996); however, this technology will not be covered in this chapter.

Erosion prediction was developed as a means to design site-specific farm management systems that control soil erosion to some specified level. The objective may be either to protect the productive soil resource base or to prevent off-farm damage, or both. The level and means of control may be different for the two objectives. For example, buffer strips and detention ponds may be used to prevent sediment from causing off-site damage, even though the erosion may be seriously degrading the soil. Degradation of the resource base may be the more serious problem, although rates of soil renewal are largely unresearched. This chapter will deal primarily with on-site damage.

Even on soils of the NWRR formed in deep loess, erosion has a detrimental impact. Nearly all soils before tillage had a developed profile, with a productive darker, high organic matter topsoil layer overlaying lighter-color, subsurface horizons that are much less productive. Erosion by either water or wind, or movement by tillage equipment on steep slopes, removes the topsoil and creates a less desirable environment for growing a crop. This chapter will deal with impacts of erosion on soil physical properties and productivity.

There have been a number of studies of the effect of erosion on productivity. This research activity reached a high level of national interest in the late 1970s and early 1980s. Two major symposia resulted: one sponsored by the American Society of Agricultural Engineers (ASAE, 1985) and one by ASA–CSSA–SSSA (Follett and Stewart, 1985).

Linking erosion rates with soil productivity is a difficult task. While it is easy to see and document the results of excessive erosion that has occurred over a long period of time, linking erosion rates to productivity at early stages after a virgin soil has been tilled or when topsoil is quite deep is more difficult. When soil has a developed profile with adequate topsoil, high erosion rates may cause little or no productivity change. Continued for 50 to 100 years, however, these same erosion rates will lead to severe soil degradation and loss of productivity. Also, technological advances during the past 40 or 50 years have compensated for much of the topsoil loss so that the results of long-term soil erosion are frequently not apparent.

3.2 PROGRESS IN EROSION PREDICTION

3.2.1 BACKGROUND

A national program on soil erosion research was undertaken in the early 1930s. Originally, ten erosion research stations were established, with more added later. Only one of the original ten, the Palouse Conservation Field Station at Pullman, Washington, was located west of the Rocky Mountains. At each station, runoff plots were established so that treatments could be compared. The control treatment was a continuous-tilled bare fallow. Data were analyzed and treatments compared and reported in technical bulletins specific to each station. For example, 10-year results from the Palouse Conservation Field Station were reported by Horner and others (Horner et al., 1944). There was no developed erosion prediction technology through which the data could be extended to apply to conditions at other locations. Later, the data were assembled and normalized so that data from all stations could be interpreted and results generalized.

Topographic effects were considered in some projects. Zingg (1940) conducted research on the effect of degree of slope and slope length on soil loss and, using his data and that of other researchers, developed the relationship:

$$x = Cs^{1.4}\lambda^{1.6} \qquad\qquad (3.1)$$

where x = total soil loss, weight units
s = land slope, %
λ = slope length, ft.
C = constant of variation

If erosion were calculated as loss per unit area, the length exponent would become 0.6.

Smith (1941) added crop and supporting practice factors to the equation to give

$$A = Cs^{7/5}\lambda^{3/5}P \qquad\qquad (3.2)$$

where A = soil loss per unit area
C = factor for weather, soil and crop management
s = land slope, %
λ = slope length, ft.
P = supporting practice factor

Smith used this relationship to develop a method for selecting conservation practices for Shelby and similar soils in the Midwest.

A few years later, a workshop resulted in the development of a relationship that became known as the Musgrave equation, named for the workshop leader. The Musgrave equation (Musgrave, 1947) was multiplicative in form, similar to Equations 3.1 and 3.2, and included factors for rainfall, slope steepness and length, soil characteristics, and vegetal cover effects. Erosion was calculated as inches of soil lost per year. The slope steepness and length exponents were reduced to 1.35 and 0.35, respectively, as compared with the Zingg recommendations of 1.4 and 0.6.

Smith and Whitt (1948) presented a relationship for soils of Missouri wherein soil loss was calculated relative to a base condition of a 90-ft.-long plot on a 3% slope. The equation is

$$A = CSLKP \qquad\qquad (3.3)$$

where A = average annual soil loss per unit area
C = average annual soil loss on a 3% slope, 90 ft. long
S = dimensionless multiplier for slope steepness
L = dimensionless multiplier for slope length
K = dimensionless multiplier for soil group
P = dimensionless multiplier for supporting practice

A rainfall factor was discussed as necessary to make the equation useful over a wider area. Other prediction advances occurred in the late 1940s and early 1950s (Lloyd and Eley, 1952; Van Doren and Bartelli, 1956).

In 1954, the National Runoff and Soil Loss Data Center was established at Purdue University. The center was responsible for collecting and analyzing the runoff and erosion data from the various studies throughout the U.S. The product of this massive effort was the universal soil loss equation (USLE), published in complete form in 1965 (Wischmeier and Smith, 1965). The USLE is written as

$$A = RKLSCP \tag{3.4}$$

where A = the predicted soil loss per unit area

R = rainfall erosivity factor, the number of erosion-index units in a normal year's rain (the erosion index is a measure of the erosive force of specific rainfall)

K = soil erodibility factor, the erosion rate per unit of erosion index for a specific soil in cultivated continuous fallow, on a 9% slope 72.6 ft. long

L = slope length factor, the ratio of soil loss from the field slope length to that from a 72.6-ft. length on the same soil type and gradient

S = slope gradient factor, the ratio of soil loss from the field gradient to that from a 9% slope

C = cropping-management factor, the ratio of soil loss from a field with specified cropping and management to that from the fallow condition on which the factor K is evaluated

P = erosion-control practice factor, the ratio of soil loss with contouring, strip-cropping, or terracing to that with straight-row farming, up-and-down slope

Maps of the rainfall erosivity factor R were developed for the eastern U.S., ending east of the Rocky Mountains. Data from the Palouse Conservation Field Station were not used in development of the USLE. Soil loss comparisons were worked up by crop stage, level of productivity, and quantity of residue after tillage at each crop stage. Wischmeier (1973) suggested a subfactor approach to account for the effect of surface cover, crop canopy, and surface roughness. This made it possible to generalize the effects of tillage and crop growth. The concept was adopted by many erosion modelers (Laflen et al., 1985).

The USLE was revised in 1978 (Wischmeier and Smith, 1978). Among other changes, the rainfall factor was renamed the rainfall and runoff erosivity factor to acknowledge the influence of runoff in the rill erosion process. Maps based on precipitation kinetic energy and intensity were extended to cover the western U.S., and an adjustment to account for thawing soil and snowmelt was offered for use in areas of the NWRR, where erosion is dominated by winter hydrology. The effect of slope steepness was modified slightly to lower the steepness factor, S for steeper slopes.

The USLE is widely used for farm conservation planning, designing practices to control erosion and sediment production from construction sites, highway cuts and fills and to select practices for mine spoil areas. Currently, the USLE is perhaps the most widely used erosion prediction technology in the world.

The revised universal soil loss equation (RUSLE) is the latest version of the USLE (Renard et al., 1997). It is computerized and calculates soil loss by 15-day (semimonthly) increments. It is more flexible and can consider a wider range of crop managements than the USLE. It is an empirical relationship and, for satisfactory performance, must be regionalized for the unique climatic conditions of the NWRR. Much, but not all, of this regionalization has been completed.

In contrast to the USLE and RUSLE, which estimate only long-term average soil erosion, the water erosion prediction project (WEPP) model (Foster and Lane, 1987) computes runoff, erosion, and sediment delivery on an individual event basis. WEPP is more process based than the USLE or RUSLE and requires substantially more information for operation. It is intended to eventually replace the RUSLE. WEPP is available in hill slope and watershed versions. The hill slope version uses a single slope profile much as the USLE or RUSLE. The watershed version contains additional components that deal with channel hydrology and hydraulics, channel erosion, and impoundments (Flanagan and Nearing, 1995). For the NWRR, WEPP includes winter routines that deal with snowmelt, effect of freezing and thawing on soil erodibility, and the transient nature of erosion processes in the region. WEPP is currently in the status of testing and validation for NWRR conditions.

3.2.2   PREDICTION ACTIVITIES IN THE NORTHWESTERN WHEAT AND RANGE REGION

Soon after development of the USLE (Wischmeier and Smith, 1965), attempts were made to apply it to the NWRR. Although erosion measurements and estimates for comparison were scanty, it was apparent that the predicted erosion rates were very low compared with measured rates, and the effect of degree of slope seemed quite large for the steep slopes of the region. Also, annual distribution of the rainfall factor, based on rainfall kinetic energy and intensity, did not properly account for erosive forces during the winter when surface runoff from rainfall and snowmelt dominates the erosion process.

3.2.2.1   1973 Adaptation of the USLE

In 1973, using several years of a small portion of the field observations of Verle Kaiser, Soil Conservation Service (SCS) State Agronomist, an interim length and steepness LS relationship was developed for the NWRR. This was reported in 1974 (McCool et al., 1974). The recommended relationships were

$$LS = \left(\frac{\lambda}{72.6}\right)^{0.3}\left(\frac{s}{9}\right)^{13} \qquad s \geq 9\% \tag{3.5}$$

and

$$LS = \left(\frac{\lambda}{72.6}\right)^{0.3}\left(0.43 + 0.30s + 0.043s^2\right)/6.613 \qquad s < 9\% \tag{3.6}$$

where LS = slope length and steepness factor relative to a 72.6-ft. slope length of
             uniform 9% slope
        λ = horizontal slope length, ft.
        s = slope steepness, %

The limited field observations were also used to develop new relationships for a
rainfall and runoff erosivity factor, R, to better account for winter effects on soil
loss. The factor, called an equivalent erosivity R_{EQ}, was based on the rainfall kinetic
energy and intensity for April through September, and an empirical relationship
between winter precipitation and measured winter erosion, corrected for soils, topography, and management. The suggested relationship was

$$R_{EQ} = 27.4 P_{(2Y, 6H)}^{2.17} + 1.5 P_{(D-M)} \tag{3.7}$$

where R_{EQ} = equivalent rainfall and runoff erosivity factor, U.S. customary
             unit
        $P_{(2Y, 6H)}$ = 2-year return interval, $6H$ duration precipitation, in.
        $P_{(D-M)}$ = December through March precipitation, in.

This relationship produced an R_{EQ} value of 35 at Pullman, Washington.

The data collected at the Palouse Conservation Field Station (PCFS) in the 1930s
and early 1940s and reported in Technical Bulletin 860 (Horner et al., 1944) were
examined for use in developing crop management factors for crops and rotations in
the NWRR. However, the data were not particularly useful because estimates and
measurements of surface cover and crop canopy were not available. In addition, the
6-ft.-wide plots were hand-tilled and were installed side-by-side with no space
between plots. Soil moisture under a fallow plot would be extracted by winter wheat
in an adjacent plot. Detailed examination of the data also indicated lack of consis-
tency in the manner certain treatments responded to runoff events. Because of
problems with these data, soil loss ratios from the Midwest and Great Plains given
in Agriculture Handbook (AH) 282 (Wischmeier and Smith, 1965) were recommended. Likewise, erodibility values from the nomograph in AH 282 were recommended. This placed all correction for the high erodibility of thawing soil in the
equivalent R_{EQ} factor. The problems encountered in developing the first-generation
adaptation of the USLE indicated almost no suitable research data were available
to use in testing and modifying the USLE or other erosion prediction models.
Research was initiated to investigate and regionalize relationships for all factors of
the USLE.

After many years without runoff plot studies, these were reestablished in the
fall of 1972 at the PCFS on the site of the 1930s plots. The initial comparisons were
quite limited in extent, and in the fall of 1976 the plots were moved to a larger area,
of slightly lower slope, that would allow a wide range of rotations and managements
with plot replication. Treatments included continuous-tilled bare fallow; winter
wheat after summer fallow, tilled; winter wheat after small grain, tilled; winter
wheat after winter wheat, no-till; winter wheat after spring peas, both till and no-till; and

rough-tilled stubble. These plot studies were continued through the winter erosion season of 1990–91.

During the years since the inception of STEEP in 1975, many of the erosion research results have been used to regionalize and improve the performance of the USLE. All of the research data also have long-term value and can be used in testing and validation of WEPP and other models. Research from 1973 through 1983 included field measurements of rill erosion. The measurements were taken from typical erosion season rill patterns along a 45- to 50-mile transect from Endicott, Washington to Troy, Idaho. The first 6 years of the data were analyzed and used to develop new LS relationships. These data were used, along with a smaller block of data collected in southeastern Idaho and north-central Oregon to develop values for a new R_{EQ} factor relationship.

3.2.2.2   1983 Adaptation of the USLE

A second-generation adaptation of the USLE to the NWRR, based on analysis of the first 6 years of field measurements, was presented in 1983 (McCool and George, 1983). The LS relationship, based on the rill data from eastern Washington and northern Idaho, was

$$LS = \left(\frac{\lambda}{72.6}\right)^{0.5}\left(\frac{\sin\theta}{\sin 5.143°}\right)^{0.7}$$ (3.8)

where LS = slope length and steepness factor relative to a 72.6-ft. slope length of
            uniform 9% (5.143°) slope.
         λ = horizontal slope length, foot
         θ = slope steepness, degrees

A relationship for winter erosivity for rill erosion only was also developed from this first 6 years of rill erosion data. A winter erosivity was calculated for each length segment on which erosion was measured by dividing measured soil loss in tons per unit area by the product of K, LS, C, and P. The value of LS was determined from Equation 3.5 or 3.6, and values of C, K, and P were assigned based on field measurements and observations. Based on runoff plot measurements, rill erosion was assumed to account for 90% of the total winter erosion. This winter erosivity for both rill and inter-rill erosion was then assumed to account for 85% of the annual erosivity. It was found that a correlation existed between mean annual equivalent erosivity R_{EQ} and MAP, and that a polynomical relationship best fit the data. The equation was

$$R_{EQ} = -72.0 + 10.3P - 0.188P^2$$ (3.9)

where R_{EQ} = equivalent rainfall and runoff erosivity factor, U.S. customary units
         P = mean annual precipitation, in.

Because this relationship would indicate a decrease of R_{EQ} with increase of precipitation above 27.4 in., it was suggested that the value of R_{EQ} would be fixed at 69 for values of P above 27.4 in.

The second-generation adaptation included a method to compute crop management factors using a subfactor approach based on material published by Wischmeier (1973). Modifications were made for crop rotations and tillage that increase infiltration and permeability. The method was programmed into a computer model that considered each tillage operation and its effect on surface and shallow (±3 in.) buried residue as well as on surface roughness. Residue decomposition with time and change in canopy cover because of growth of the seeded crop were also considered.

3.2.2.3   Adaptation of RUSLE

Development of the RUSLE was started in 1985. Analysis of field and plot erosion data was directed toward developing relationships to regionalize RUSLE for the NWRR.

By 1987, the entire 10 years (no data were collected in the spring of 1977 because of severe drought) of field rill erosion data from eastern Washington and northern Idaho had been analyzed. The data included slopes from 1.5 to 56% steepness and 62 to 660 ft. in length. The relationship for LS was

$$LS = \left(\frac{\lambda}{72.6}\right)^{0.5} \left(\frac{\sin\theta}{\sin 5.143°}\right)^{0.6} \tag{3.10}$$

where LS = slope length and steepness factor relative to a 72.6-ft. slope length of uniform 9% (5.143°) slope
λ = horizontal slope length, ft.
θ = slope steepness, degrees.

Later, because a relatively small portion of the data was collected from slope segments of less than 9% steepness, and use of Equation 3.10 for slopes of 5% or less produces relatively large LS values that appear unrealistic, it was recommended (McCool et al., 1993) that for slopes less than 9% the following equation, developed from rainfall simulator data on plots of 0.1 to 3% steepness (McCool et al., 1987), be used:

$$LS = \left(\frac{\lambda}{72.6}\right)^{0.5} (10.8\sin\theta + 0.03) \qquad s < 9\% \tag{3.11}$$

Two approaches were considered in developing the current equivalent erosivity factor R_{EQ} for the NWRR. One approach was to use the data from the field rill erosion measurements.

Calculation of the equivalent R_{EQ} involves, for each measured rill pattern on each field, calculating a winter rill equivalent erosivity, R_{WR}

$$R_{WR} = A/(KLSCP) \qquad\qquad (3.12)$$

where R_{WR} = equivalent winter rill erosivity
        A = measured rill erosion
        K = soil erodibility
        L = slope length factor
        S = slope steepness factor
        C = crop management factor
        P = supporting practice factor

The A is a measured quantity; K is obtained from the erodibility nomograph; L and S are calculated from the specific slope data; C is calculated from field estimates of surface cover, canopy cover, prior land-use effects, surface roughness, and an antecedent moisture correction; and P is estimated from field observations. The R_{WR} at a measured site is then increased for inter-rill erosion, and again for spring, summer, and fall erosivity. The resulting annual equivalent erosivity factor R_{EQ}, since values are available from many fields, can be correlated with annual precipitation and a general relationship developed. A linear relationship is indicated by the data. The values of R_{WR} at a site, and hence the entire relationship for R_{EQ}, are dependent on the C factor; any change in the relationships for the subfactors of the C factor can result in a change in the R_{EQ} relationship. The relationships for the effect of the subfactors of C are not yet stable. Thus, this approach results in the potential for continuous changes in the equivalent R_{EQ} value at a given location as relationships for subfactors of C are refined and revised.

An alternative approach for developing a relationship between an equivalent R_{EQ} and precipitation or some other climatic parameter was to use the continuous-tilled bare fallow (CBF) plots at the PCFS as benchmark values. The two plots were installed, one in the fall of 1978 and the other in the fall of 1979. After the first 3 years, it became obvious the plots were changing with time in their response to potential erosion events; their response, as compared with other rotations, stabilized in the third year. Thus, the first 2 years of data from each were deleted.

This approach removed all reliance on C factors, as the C factor for the CBF condition is 1.0. Also, since the plot had no ridges and the few marks left by the tillage (a large rototiller) were raked smooth, the P factor was 1.0. Adjustment for K and LS was necessary. The LS values for the 73-ft. plot on 21.5% slope, and the 73-ft. plot on 15.6% slope were 1.67 and 1.38, respectively, based on Equation 3.10. The nomograph K was 0.32 for the Palouse silt loam on both plots. The calculated equivalent R_{EQ} value for the plots was 143 and 90, giving an average value of 115.

A linear relationship between equivalent R_{EQ} and MAP was assumed as indicated by the field rill erosion data. The MAP value at Pullman is 21.1 in. A minimum R value of 10, based on erosion index calculations, has been assumed for lower rainfall areas of central Washington. It was further assumed that the R_{EQ} would be applicable above MAP of 7.5 in., which approximates the lowest precipitation at which nonirrigated winter wheat can be grown in the NWRR. Thus, another benchmark

R_{EQ} value of 10 was fixed at annual precipitation of 7.5 in. The resulting relationship for R_{EQ} is

$$R_{EQ} = -48 + 7.8P \qquad P \geq 7.5 \tag{3.13}$$

where R_{EQ} = equivalent annual erosivity factor, U.S. customary units
         P = mean annual precipitation, in.

Erodibility K values currently in use in the NWRR are developed from the soil erodibility nomograph (Wischmeier et al., 1971). In the eastern areas of the U.S., a time-variable K is used. The relationship for variation with time is based on the annual R value calculated from erosion index units. The relationship was not reliable for the western U.S. It is recognized that erodibility varies with winter freezing and thawing, and moisture increase during the winter and decrease during the summer growing season; however, at this time it is not possible to address this matter. In the NWRR, all freeze/thaw effects on erodibility are absorbed by the equivalent R_{EQ} factor.

Crop management factors C are average values calculated over the period of a crop rotation. The soil loss ratio (SLR, ratio of soil loss under a given treatment to that from CBF) for a given time period is multiplied times the fraction of annual erosivity during that period. These products are summed for the duration of a rotation and divided by the number of years in the rotation to give the rotational C factor.

Following the procedure proposed by Wischmeier (1973) and subsequent developments, the relationship in RUSLE for the SLR is

$$SLR = PLU \cdot CC \cdot SC \cdot SR \tag{3.14}$$

where   SLR = soil loss ratio for the given treatment
         PLU = prior land-use subfactor
         CC = crop canopy subfactor
         SC = surface cover subfactor
         SR = surface roughness subfactor

Detailed data were not available from the runoff plots at the PCFS to enable testing of relationships for most of the subfactors of Equation 3.14, but comparison of runoff plot results with crop stage C values in AH 282 indicated erosion differences between wheat after fallow and wheat after small grain were not properly accounted for in AH 282. Therefore, based on field observations, when the soil profile moisture content was high vs. low, a soil moisture subfactor, SM, was added to Equation 3.14.

Thus, Equation 3.14 becomes

$$SLR = PLU \cdot CC \cdot SC \cdot SR \cdot SM \tag{3.15}$$

where   SM = soil moisture subfactor

Values of the soil moisture subfactor are based on different runoff and erosion rates between winter wheat seeded following winter wheat, summer fallow, or spring crop. This is intended to account for the effect on infiltration and permeability of whether the rooting depth is at wilting point, near-field capacity, or at some intermediate level in the fall prior to the winter erosion season, as well as the rate at which the soil moisture is replenished during the winter precipitation season.

Values of SM vary between 0.0 and 1.0, based on whether the soil profile is at wilting point or field capacity. When the soil profile is at or near field capacity, SM is 1.0. When the profile is near wilting point to a 6-ft. depth, the SM value is 0, indicating that no runoff and erosion are expected. This assumes that infiltration is not limited by surface conditions. The SM value increases over the winter from October 1 through March 31 and decreases from April 1 through August 31, and is constant from September 1 through 30. Suggested replenishment-rate relationships are given in Figure 3.1. Growing season (April 1 to July 31) depletion rates for typical crops appear in Table 3.1. These relationships and values will need adjustment for shallow soil conditions or other considerations.

Most of the relationships for the subfactors of the SLR come from the eastern U.S. or are a mixture of eastern U.S. and NWRR relationships. The prior land-use subfactor, PLU, is based on surface consolidation and incorporated biomass in the near-surface layer of the soil. Data collected from an experiment at the PCFS and reported by Van Liew and Saxton (1983) were used in developing the relationship.

The crop canopy subfactor, CC, is based on interception of raindrops by crop canopy. The relationship is assumed to be linear. When crop canopy is of very low height, CC has a value of 1.0 when there is no cover and a value of 0.0 when there is 100% canopy cover. The benefit is reduced for taller plants and trees that intercept raindrops that then drip to the soil from some height. Random distribution of canopy is assumed in the relationship. Preliminary data from recent studies at the PCFS and on producer's fields indicate that when the canopy is not random but is in rows at the bottom of a furrow, a given mean percent cover has much more effect than previously assumed. In fact, on plots with up-and-down slope 10-in. row spacing, and low height canopy cover of 0 to 40%, a fitted linear relationship projects negligible soil loss would be reached at less than 60% average canopy cover. This relationship is tentative and requires further testing before use.

TABLE 3.1
Growing Season Soil Moisture Depletion Rates for Use in Calculating Soil Moisture Subfactor for Use in the NWRR

Crop	Depletion Rate
Winter wheat and other deep-rooted crops	1.00
Spring wheat and barley	0.75
Spring peas and lentils	0.67
Shallow-rooted crops	0.50
Summer fallow	0.00

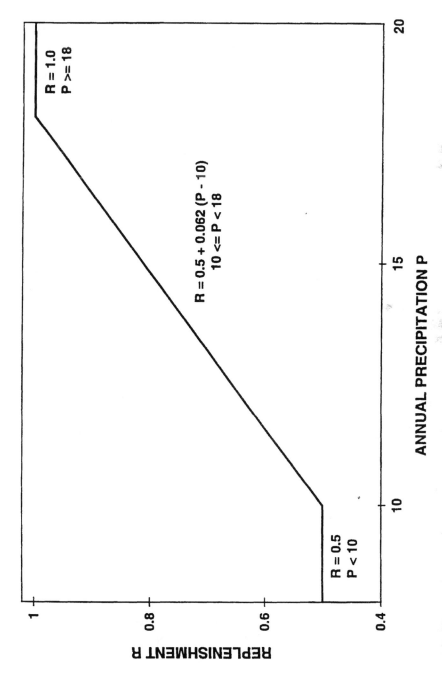

FIGURE 3.1 Fraction of soil moisture replenished over winter vs. annual precipitation for calculation of a soil moisture subfactor for the NWRR.

The surface roughness subfactor, SR, is based on data from the eastern U.S., but appears to give reasonable results in the NWRR. In parts of the NWRR, clods appear to be very important in controlling erosion because those clods formed when very dry soil is tilled after harvest are quite resistant to breaking down over the winter. When soils are tilled at high moisture content, the clods are smaller and break down more easily over the winter. The SR value is based on the random roughness of the soil surface. Field observations indicate that a random roughness value of 2.5 in. is highly effective in controlling erosion.

Data from the PCFS plots indicated the commonly used eastern U.S. (Laflen et al., 1985) relationship for surface cover subfactor, SC:

$$SC = \exp(-bM) \tag{3.16}$$

where   SC = surface cover subfactor
        b = fitted coefficient
       M = percentage surface cover

with a b value of 0.035 did not show adequate credit for the surface cover. Thus, a new value of b was developed from the PCFS plot data. This was done by computing the SLR value for each treatment, and computing values for the subfactors PLU, CC, SR, and SM. Because the seeding and tillage operations were on-contour, it was also necessary to compute contour support practice factor, PC, values as well. The product of the PLU, CC, SR, SM, and PC values was divided into the SLR to give a value of SC for each treatment. The SC values were fit to give a value of $b = 0.05$ in Equation 3.16. This value may change slightly with the addition of new data. The b value of 0.05 indicates a greater benefit of surface cover for primarily rill erosion that is consistent with data from other research. There may also be interaction with the low precipitation intensity and generally low runoff rates that allow even small pieces of surface residue to remain in place without being floated off as with higher rates of runoff.

Most erosion in the NWRR occurs during the lengthy winter erosion season and a brief period after spring crop seeding. Assuming that most farm plans rely heavily on crop management to control erosion during these periods, an analysis of the sensitivity of soil loss to roots and incorporated residue, surface cover, crop canopy and height, surface roughness and soil moisture, based on current relationships, is of interest. This analysis does not indicate which subfactor is most important, but rather how a change in one element affects the predicted erosion. In the early winter after seeding winter wheat, conditions of 20% canopy cover with a 3-in. crop height, surface cover of 25%, random roughness of 0.5 in., incorporated residue of 4000 lb acre^{-1} and root mass of 300 lb acre^{-1} in the upper 4 in. of the soil following a medium-to shallow-rooted spring crop would not be unusual in the annual cropping zone. A sensitivity analysis based on current relationships for the NWRR used in RUSLE indicates that small changes in surface cover and incorporated surface biomass result in greatest percent change in soil erosion. Changes in root mass, crop canopy, and surface roughness have much less effect. At the low crop height assumed, changes in crop height have almost no effect on soil erosion. The effect of changing the parameters from the assumed base values is shown in Figure 3.2.

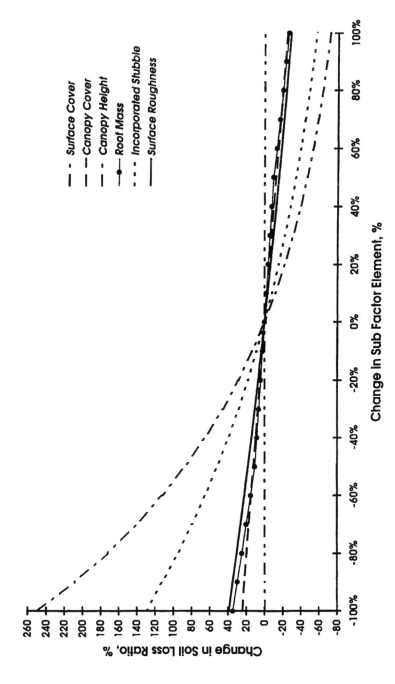

FIGURE 3.2 Sensitivity of soil loss ratio to subfactor elements.

The conservation support practice factor, *P* values are developed using technology from the Chemicals, Runoff, and Erosion from Agricultural Management Systems (CREAMS) model (Knisel, 1980), and have been regionalized for the NWRR by using runoff curve number values and other parameters specific to the low-intensity precipitation, frozen soil conditions of the region. Contouring benefits are dependent on precipitation characteristics, ridge height, furrow grade, slope length and steepness, and soil surface condition. Strip-cropping practice in the NWRR differs from that in the eastern U.S. in that in the eastern U.S. grass strips alternate with strips of seeded crop, whereas in the NWRR strips of seeded winter crop alternate with standing or rough-tilled stubble. Under conditions of very rough-tillage, the tilled stubble will infiltrate runoff from the seeded crop strip. Grass, if kept clipped or harvested, may have an infiltration rate similar to the seeded winter crop and hence protect itself but not the next strip downslope. Thus, rough-tilled stubble is a desirable winter condition for the unseeded strip. Standing stubble will collect snow and prevent soil freezing, but any rills left from the previous erosion season will pass runoff from seeded strip to seeded strip, negating much of the benefit of the strip system. Strip-cropping in annual cropping areas has a period of vulnerability in the spring after the spring crop is seeded but before growth is adequate for protection. Strip crop benefits are dependent on the same factors as contouring, as well as strip width.

The RUSLE program requires three databases, one for climate (CITYLIST), one for vegetation (CROPLIST), and one for field operations (OPLIST). The model cannot produce reliable output without the appropriate databases. Each record in the climate database includes an index number, a city or place name, annual *R* factor value (also an R_{EQ} value for the frozen soil areas of the NWRR), 10-year frequency storm EI, mean monthly precipitation and temperature, as well as 24 entries for semimonthly *R%*.

The vegetation database includes for each crop a name, yield, and residue production, decomposition parameter for above- and belowground biomass, and residue weight per unit area at 30, 60, and 90% residue cover of the soil surface. The remainder of the individual crop record includes, by 15-day interval, root mass in the upper 4 in. of soil, percent canopy cover, and water drop fall height. Most of the crop databases in use in the NWRR were developed using the cereal canopy cover model crops for RUSLE (CRP4RUSL), developed by Rickman (Rickman et al., 1994). The model provides estimates for root mass, crop canopy, and drip height required at 15-day intervals by RUSLE.

The field operation database includes effects of tillage implements as well as grain harvest, hay cutting, or start of regrowth. Each record includes an operation name, percent of soil surface disturbed, random roughness after operation, percent of before operation mass or cover retained on the surface, and depth of tillage.

3.2.2.4 WEPP

Erosion researchers in the NWRR have been active in the WEPP project since its inception in 1985 (Foster and Lane, 1987). WEPP is a process-based computer model and requires winter subroutines be developed to predict snowmelt, soil freezing and thawing, and effects of freeze/thaw on infiltration and runoff as well as on soil erodibility (Flanagan and Nearing, 1995). Two soils in the NWRR were selected for inclusion in the series of simulated rainfall studies for WEPP. Sites with Palouse and Nansene (similar to Walla Walla silt loam) soils were located, prepared in the fall of 1986, and tested in the summer of 1987. Data from these sites are included in the WEPP cropland database (Elliott et al., 1989).

A series of studies on the effect of freezing and thawing on soil strength and erodibility were conducted in laboratory and field settings at the PCFS (Formanek et al., 1984; Van Klaveren and McCool, 1987; Kok and McCool, 1990). These results have not as yet been included in the winter routines of WEPP.

Testing of a version of WEPP that includes winter routines currently is under way using data collected from runoff plot studies at the PCFS.

3.2.2.5   Validation Activities

Data to validate runoff and erosion models are expensive and difficult to collect. Projects are frequently set up for much too short a time to include a range of weather conditions. Thus, rainfall simulators are widely used as a technique to replace natural rainfall plots. Yet this method has severe disadvantages in a winter erosion area. Inasmuch as many major events occur as the soil is thawing after a freezing period, the devices must generate rainfall under near-freezing conditions. Frozen soil conditions are the signal to prepare for simulator runs, but the ideal conditions for simulation are short lived and may occur on weekends or holidays. To date, this technique has not generated large amounts of winter data in spite of the development of two simulators that mimic the characteristics of natural rainfall (Bubenzer et al., 1985; Williams et al., 1998).

Data available for validation of runoff and erosion models include published data from the 1930s from plots at the PCFS (Horner et al., 1944) and plot data collected in the 1970s and 1980s from the PCFS and Kirk site near the Columbia Plateau Conservation Field Station near Pendleton, Oregon. Cesium-137 studies have also been conducted in which movement of fallout-based cesium-137 was tracked in two studies (Busacca et al., 1993; Montgomery et al., 1997). The cesium-137 studies are useful to illustrate magnitude of soil movement of combined water erosion and tillage translocation. The data are of limited use in model validation because cesium-137 measures both tillage and water erosion effects and there is rarely a complete record of tillage operations and resulting surface cover and random roughness unless the area has been under study since the fallout of the early 1960s nuclear tests.

3.3   DAMAGE CAUSED BY EROSION

3.3.1   Effect of Erosion on Topsoil Depth and Other Soil Properties

Reasons widely cited for decline in productivity of soils with progressive erosion are (1) loss of soil organic matter, (2) decrease in volume of rooting with associated reduction of plant-available water and nutrients, (3) reduced water infiltration,

(4) reduced tilth, and (5) reduced numbers of beneficial soil biota (Langdale and Schrader, 1982; Schertz, 1983; Pimentel et al., 1995). All of these changes have been observed under cultivation and erosion in the NWRR and STEEP research has helped to quantify these changes and their impact on crop yields.

The basic hill shapes, soil profile properties, and soil-landscape patterns differ from one agroclimatic zone to another, so the impacts of water erosion differ as well. Two examples from STEEP research, originally reported by Busacca and Montgomery (1992), illustrate the impacts of accelerated water erosion on soil physical properties across NWRR farmed landscapes. The first site is in western Whitman County, Washington where there is about 14 in. of MAP. Soils there have moderate amounts of humus in the topsoils and lime-silica accumulations in subsoil horizons. The second site is in eastern Whitman County and is typical of areas of the NWRR that receive about 18 to 23 in. of MAP. In this agroclimatic zone, the topsoils have more organic matter and clay-rich argillic horizons restrict and control subsurface water flow and rooting.

At the first site, several traverses from north to south across a typical farm field established that soils of the gently sloping summit are generally about 4 ft. to a restrictive calcic or duripan subsoil horizon (Figure 3.3a). Deep, fertile soils like these are in the Athena, Oliphant, Walla Walla, and similar soil series. Although organic carbon contents of the tillage layer of such soils may have been up to 2.0 to 2.5% in native condition, the measured levels today are only 0.3 to 1.7% (Figure 3.3b). This decreases nutrient storage and availability and reduces aggregate stability, which may increase the severity of future soil losses to water erosion.

The soils are very thin on the south-facing backslopes and on eroded midslope knobs. These are places in the field where water and tillage erosion have had great impact on soil productivity. Calcic or cemented duripan materials are only a few inches to 1.5 ft. beneath the surface. Their depth in most cases marks the limit of useful plant rooting and water storage for crops. No bulk density measurements were made, but a power soil probe was able to penetrate only a few inches into the calcic subsoil. In these places, large fragments of duripan at the surface interfere with tillage operations and seedling emergence. The calcium carbonate or lime in the tillage zone, which ranges up to 5% in this field (Figure 3.3b) and up to 15 to 20% in extreme cases, also causes deficiencies of micronutrients such as Fe, Zn, B, and Mn and lowered availability of phosphorus (Pan et al., 1992). Soils like those of the Lance series in Whitman County have this shallow profile. In freshly plowed fields and on aerial photographs, whitish zones with lime at the surface are easy to spot and the area they occupy in fields can be estimated by remote sensing methods (Frazier and Busacca, 1987).

The second STEEP research site reported by Busacca and Montgomery (1992) is in the classical Palouse landscape, where steeply sloping hills with complex shapes are formed in deep loess. This classical landscape is seen in eastern Whitman and southern Spokane Counties, Washington and western Latah County, Idaho. Steeply sloping hills occur in the same agroclimatic zone in other parts of the NWRR, such as in Columbia and Garfield Counties, Washington and in Umatilla County, Oregon, but in these places loess is thin and basalt bedrock is often within a few feet of the surface.

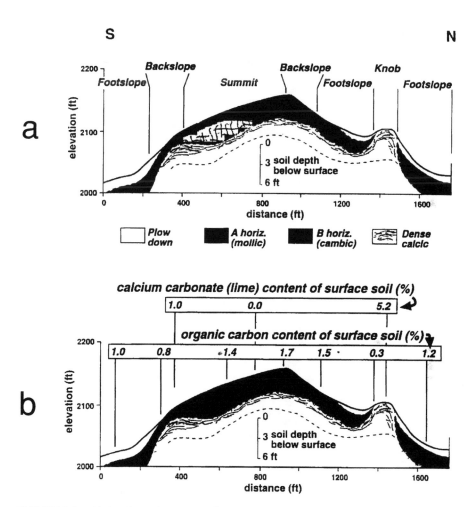

FIGURE 3.3   (a) A schematic cross section of a site in the 15-in. MAP zone showing soil profile features and named landscape positions. Three different scales are used: one for the horizontal distance across the hill, a second for the elevation of the hill, and a third for the depth of the soil horizons beneath the land surface. (b) A schematic cross section of the same site shows the amount of calcium carbonate (lime) and of organic carbon in the top 4 in. of the soil at selected points (indicated by vertical lines) along the cross section.

At this site, a series of ten backhoe trenches was excavated across a hill summit, from which a cross section of the soil features was drawn (Figure 3.4; Busacca and Montgomery, 1992), much as was done first in 1953 by Lotspeich and Smith. The hill at this site has a narrow summit and extensive backslope and footslope areas. The fertile mantle of young loess consists of a dark topsoil or A horizon and a brown, well-structured subsoil or cambic horizon (Figure 3.4). In the south-facing footslope and toeslope positions, the soil profile is more than 4 ft. deep, including a thick wedge of surface soil material transported by tillage to the base of the footslope (McCool et al., 1997). The soils on the equivalent north-facing positions

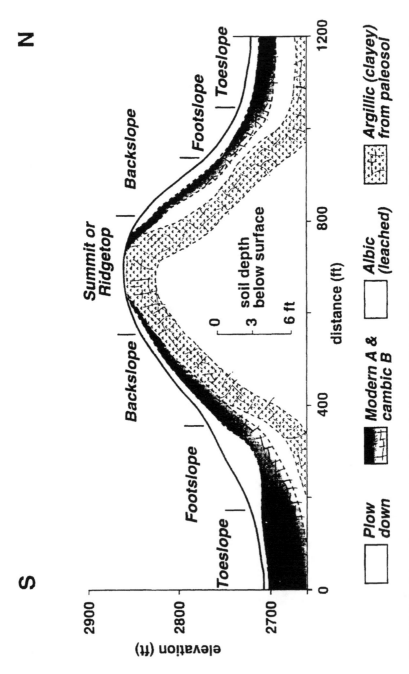

FIGURE 3.4 Schematic cross section of a site in the 15 to 23 in. MAP zone showing soil profile features and landscape positions. Three different scales are used: one for the horizontal distance across the hill, a second for the elevation of the hill, and a third for the depth of the soil horizons beneath the land surface.

NaOAc extractable P (ppm)

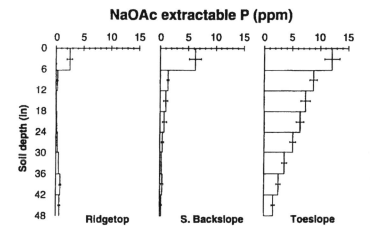

FIGURE 3.5 Extractable phosphorus by position and depth in an eastern Palouse landscape. (From Pan, W.L. and Hopkins, A.G., *Plant Soil*, 135, 9–19, 1991. With permission.)

are not quite as deep; however, at other sites, the north-facing soils typically are deeper than the south-facing ones. The surface soil thins in the upslope direction and is less than 2 ft. thick on the south-facing backslope. Immediately underlying the soil in the modern loess in this position is a dense, clay-rich paleosol argillic horizon.

The shallowest soil is on the summit, where all of the modern loess has been lost due to accelerated water and tillage erosion, although it was thinner there at the start of farming because of geologic erosion. On the summit, the plow zone is in the dense argillic horizon. Soils with profiles like those of the upper backslope and summit in the Pullman area are in the Garfield, Naff, and Staley series. These soils tend to shed water because of their landscape position and have low infiltration rates because of poor soil structure (Pierson and Mulla, 1990). This, in turn, increases water runoff to lower slope positions and decreases plant-available water on shallow soil sites. They can crust severely, which will reduce seedling emergence. Deficiencies of nutrients such as phosphorus occur on these exposed clayey subsoils (Figure 3.5; Pan and Hopkins, 1991; Pan et al., 1992). The dramatic differences in organic carbon content across the eroded landscape of Figure 3.4 can lead to nitrogen mineralization rates on summits that are less than one tenth the rates on backslopes or footslopes (Fiez et al., 1995).

In this precipitation zone, an *albic* (leached and white) subsoil horizon commonly overlies the argillic horizon on backslope, footslope, and toeslope positions of north-facing slopes (Rieger and Smith, 1955). The depth, landscape distribution, and thickness of the albic feature indicates the probability of seasonal wetness and of management problems associated with wetness (McDaniel and Falen, 1994), low soil temperatures, and losses of nitrogen due to denitrification (Pan et al., 1992).

In other STEEP research, changes in soil properties due to the effects of annual cropping and water erosion have been estimated by comparing a Palouse native grassland site to a nearby farmed site (Rodman, 1988; Busacca and Montgomery,

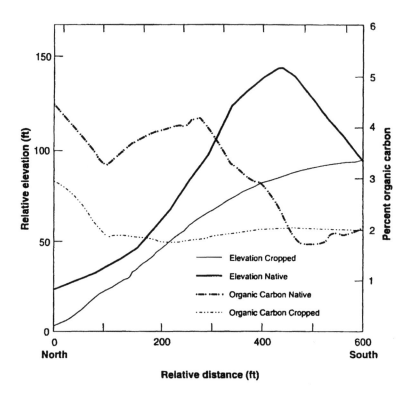

FIGURE 3.6   Two hill slope cross sections for adjacent native and cropped hills in the 18 to 23 in. MAP zone. Solid lines are the hill outlines, with relative elevation shown on the left axis, 1 m equals 3.3 ft. Contents of organic carbon in the upper 4 in. of the soils is shown by the dashed lines. Scale of organic carbon percentages is on the right axis. (From Busacca, A.J. and Montgomery, J.A., in *Precision Farming for Profit and Conservation, 10th Inland Northwest Conservation Farming Conference Proceedings*, Washington State University, Pullman, 1992, 8–18. With permission. Data from Rodman, 1988.)

1992) Organic carbon in the native site ranges from about 2% on the summit to almost 4.5% on the north-facing footslope (Figure 3.6). The large differences in organic carbon between the native and the cropped site at comparable landscape positions (2 to 3 times) are due mostly to losses by erosion and continuous cropping (Rodman, 1988).

In the eastern Palouse, properties such as clay content and bulk density of the tillage zone follow from the erosion history; that is, clay content and bulk density are high where the soils are shallow and eroded. The clay content of the surface soil ranged from 18 to 28% and as high as 38% in one STEEP research watershed (Busacca and Montgomery, 1992; Montgomery, 1993). Higher clay contents were found in two different settings—on summits and knobs where clayey argillic horizons are exposed and in the drainage ways where some of the eroded subsoil material is redeposited. These clay-rich eroded materials can cover areas that have otherwise intact and fertile topsoils. Bulk density of surface soils can indicate serious compaction problems or can indicate that dense subsoils are at the surface. The bulk density of the surface soils in the watershed of Busacca and Montgomery (1992) ranged from 1.1 to 1.60 g/cm^3. The densest materials were argillic subsoil materials displaced downslope of summits and eroded knobs. Bulk densities higher than about 1.2 g/cm^3 may strongly limit root penetration and proliferation in the degraded parts of the field landscape and may decrease water infiltration.

3.3.2 Topsoil Depth and Crop Productivity

Soil and crop performance are affected strongly by soil erosion because erosion reduces the effective rooting zone and reduces the beneficial properties of the soil remaining in the rooting zone. Pioneering soil conservationist Verle Kaiser was perhaps the first to suggest that accelerated soil erosion in the Palouse diminishes soil productive potential (Kaiser, 1967). He correctly deduced that the economic and environmental cost of soil erosion is masked by the upward trend of yields over time due to technological improvements in farming. Research in the NWRR starting in the 1950s and continuing under STEEP has shown that, as a simple approximation, winter wheat yields across the Palouse decrease with decreasing topsoil depth and that yields are reduced more strongly on soils with initially thin topsoils than on those with thick topsoils (Figure 3.7; Pawson, 1961; Wetter, 1977; Busacca et al., 1985; Young et al., 1985; Fosberg et al., 1987). Pawson (1961) apparently was the first to establish the magnitude of yield loss due to declining topsoil depth by making more than 800 measurements in eastern Palouse farmer's fields in 1952 and 1953 (bottom curve, Figure 3.7). The data of Wetter (1977) from 89 observations of farmer's fields in the same area during 1970 to 1975 was used to estimate the top curve in Figure 3.7. Similar topsoil depth–crop yield relationships have been established by recent work in other areas, for example, in Canada (Verity and Anderson, 1990; Larney et al., 1995).

Papendick et al. (1985) estimated that the loss of productivity has been least severe on soils having moderately low erosion rates [<6.7 tons/acre/year] and thick topsoils (>12 in.) and most severe on soils having high erosion rates [>17.8 ton (acre/year)] and thin topsoils (<6 in.). One simple estimate based on the data in Papendick et al. (1985) is that yield of winter wheat declined by 0.41 bu/acre per inch of topsoil loss for soils with topsoil thickness of 21.6 in. and that yield declined 3.5 bu/acre for soils with topsoil thickness of 2.8 in. (Busacca et al., 1985).

Bramble-Brodahl et al. (1985) and Fosberg et al. (1987), in STEEP-sponsored research, evaluated wheat yield response to soil loss by measuring differences in yield with differences in existing topsoil depth for the same soil series at different locations within fields. Slope, aspect, crop vigor, and weed competition were similar at measurement points. Both linear and nonlinear regression models were examined to relate yield to thickness of the mollic epipedon (the taxonomic term for topsoil) or to the depth to an argillic (clay-enriched) subsoil horizon if one existed in the soil being considered.

Results for the Palouse series soils, which have a less restrictive cambic subsoil horizon, and the Naff series soils, which have a restrictive argillic horizon, illustrate the general relationships between yield and topsoil depth established by Bramble-

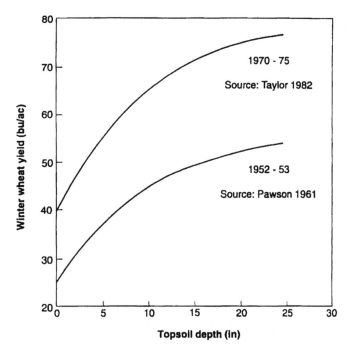

FIGURE 3.7 Comparison of winter wheat yield–topsoil depth relationships from the 1950s and the 1970s, eastern Whitman County, Washington. (From Young, D.L., et al., in *Erosion and Soil Productivity, Proceedings of the National Symposium on Erosion and Soil Productivity,* New Orleans, 10–11 December 1984, American Society of Agricultural Engineers, St. Joseph, MI, 1985, 130–142. With permission.)

Brodahl et al. (1985) and Fosberg et al. (1987). For the 1982 crop year, the yield–topsoil thickness relationship for the Palouse soil was linear with an R^2 of 0.43 (Figure 3.8a). This model described a yield decrease of 1.4 bu/acre/in. of topsoil loss from a base yield of 50.6 bu/acre at zero topsoil (the y intercept of the model; Bramble-Brodahl et al., 1985). For the 1982 crop year, a linear model that related yield to depth to the argillic horizon (depth to Bt; Figure 3.8b) for the Naff soil yielded a higher R^2 (0.50) than did a linear model that used topsoil thickness ($R^2 = 0.18$), indicating that the depth to this water- and root-inhibiting subsoil horizon has biological significance (Bramble-Brodahl et al., 1985). This model described a yield decrease of 1.8 bu/acre/in. of topsoil loss from a base yield of 41.6 bu/acre at zero topsoil.

Busacca et al. (1985) used a Mitscherlich–Spillman equation derived by Hoag (1984) to interpret 213 wheat yield–topsoil depth measurements collected by the Soil Conservation Service from farmer-cooperators' fields between 1966 and 1980. From this equation they derived the responses of Palouse, Thatuna, and Naff soils to diminishing topsoil depth (Figure 3.9). Higher yields are predicted at every topsoil depth for the Palouse and Thatuna soils than for the Naff soils. The response function is more steeply sloping for the Thatuna series soils than for either the Naff or Palouse

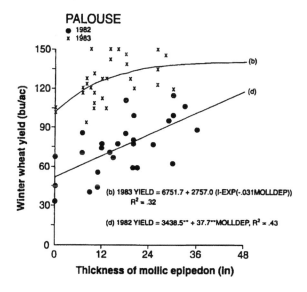

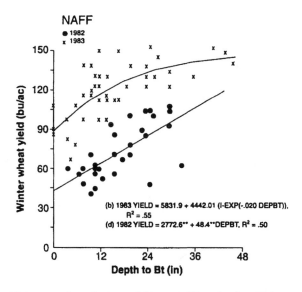

FIGURE 3.8   (a) A regression of wheat yield vs. mollic epipedon thickness for the Palouse series soil for 1982 and 1983; (b) the regression of wheat yield vs. depth to the argillic subsoil horizon for the Naff series soil for 1982 and 1983. (From Bramble-Brodahl, M., et al., in *Erosion and Soil Productivity, Proceedings of the National Symposium on Erosion and Soil Productivity*, New Orleans, 10–11 December 1984, American Society of Agricultural Engineers, St. Joseph, MI, 1985, 18–27. With permission.)

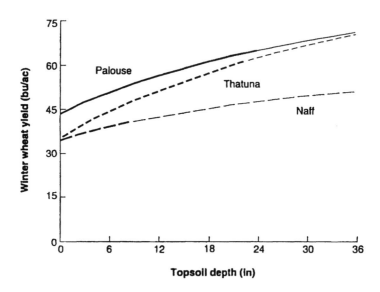

FIGURE 3.9 Topsoil depth–wheat yield response functions for the Palouse, Thatuna, and Naff series soils on land use capability class III sites. Heavier line segments represent the natural range of topsoil thicknesses for undisturbed to eroded sites of each soil; lighter line segments are extrapolations beyond this range derived from eq. 1 of Busacca et al. (1985). (With permission.)

soils. Both of these predictions have physical significance in the differing characteristics of these soils. In the first case, the cambic subsoil of the Palouse series is far more favorable for root and water penetration under any erosion condition than is the clayey argillic subsoil in the Naff series soil. The steeply sloping response function for the Thatuna soil also follows from soil characteristics. When its topsoil is deep, the Thatuna series produces wheat yields that are comparable to those of the Palouse series (Figure 3.9; Busacca et al., 1985). When its topsoil is thinned by erosion, however, yields drop to levels more characteristic of the Naff series. Palouse soils have deep topsoils and favorable cambic subsoils that offer minimal limitations to roots and water and good fertility. Thatuna soils have comparably deep topsoils but they have a dense clayey argillic horizon at 27.5 to 39.0 in. As topsoil is lost from both soils, productivity declines more markedly on the Thatuna series than on the Palouse and as erosion becomes severe on Thatuna soils, their yields approach the limiting minimum of the Naff soil with its clayey argillic horizon exposed. These results, along with those of Bramble-Brodahl et al. (1985) and Fosberg et al. (1987), provide strong evidence that soil erosion leads to yield declines of winter wheat, that the response varies by soil series, and that the yield response functions generally are consistent with soil profile properties.

A number of STEEP research studies have expressed this in economic terms of decreased crop yields today and decreased yield potential in the future (Busacca et al., 1985; Walker et al., 1987; Fosberg et al., 1987). Bramble-Brodahl et al. (1985) used a damage model to estimate the annual cost per acre to be $11.60 for the

Palouse soil and $16.30 for the Naff soil in lost income due to decreasing topsoil depth. These monetary losses were based on the assumption of annual soil losses of 10 ton/acre, a 75-year time horizon over which damage from erosion this year impacts yields in the future, and assumptions about the price of wheat, the rate of technical progress, and others.

Young et al. (1985) demonstrated that recent agricultural technical progress has produced greater yield increases on deeper topsoils than on shallower topsoils for winter wheat grown in the Palouse (compare the curves in Figure 3.7). In simple terms, the technological advances are more effective on deeper topsoils. This means that the multiplicative effect of technological progress on yield damage from erosion leads to an underestimation of the damaging impact of erosion on future yields if this effect is not taken into account in economic models.

3.4   SUMMARY

The NWRR offers unique challenges in erosion modeling and determining the effects of erosion on productivity. Unlike other areas of the U.S., most water erosion occurs during the winter months from rainfall and snowmelt. Thawing soils with a water-restrictive frost layer are frequently involved. These thawing soils are of very low strength and erode quite easily with small amounts of rainfall on runoff.

Furthermore, the wind-deposited soils are of variable thickness over rock, water-restrictive layers are present in the soils of much of the area, slopes range from gentle to quite steep, and there are major precipitation gradients.

Significant progress has been made since the inception of STEEP in both erosion modeling and in determining the effects of erosion, primarily through decreases in topsoil depth, on productivity. During the STEEP project, regionalized relationships for factors of the USLE and RUSLE were developed and have proved reliable. Testing of WEPP is currently in progress.

Concurrently, effects of erosion on productivity were determined. Effects of erosion are not evident when topsoil depth is great. However, for shallow soils and soils that have lost large amounts of topsoil such that water and root restrictive layers are present in the water extraction zone, soil erosion has a significant impact on productivity, mostly through effects on the available water supply. No models have been developed as yet that link soil erosion with the effects of erosion on productivity.

REFERENCES

American Society of Agricultural Engineers (ASAE), 1985. Erosion and Soil Productivity, ASAE Publ. 8-85.

Austin, M.E., 1981. Land resource regions and major land resource areas of the United States, U.S. Department of Agriculture, Agricultural Handbook 296.

Berger, G.W. and A.J. Busacca, 1995. Thermoluminescence dating of late Pleistocene loess and tephra from eastern Washington and southern Oregon and implications for the eruptive history of Mount St. Helens, *J Geophys. Res.*, 100(B11), 22,361–22,374.

Bramble-Brodahl, M., M.A. Fosberg, D.J. Walker, and A.L. Falen, 1985. Changes in soil productivity related to topsoil depth on two Idaho Palouse soils, in *Erosion and Soil Productivity, Proceedings of the National Symposium on Erosion and Soil Productivity*, New Orleans, 10–11 December 1984, American Society of Agricultural Engineers, St. Joseph, MI, 18–27.

Bubenzer, G.D., M. Molnau, and D.K. McCool, 1985. Low intensity rainfall with a rotating disk simulator, *Trans. ASAE*, 28(4), 1230–1232.

Busacca, A.J., 1989. Long Quaternary record in eastern Washington, U.S.A., interpreted from multiple buried paleosols in loess, *Geoderma*, 45, 105–122.

Busacca, A.J., 1991. Loess deposits and soils of the Palouse and vicinity, in V.R. Baker et al., The Columbia Plateau, chap. 8, in R.B. Morrison, Ed., *Quaternary Non-Glacial Geology of the Conterminous United States*, Geological Society of America, *Geology of North America*, Vol. K-2, Geological Society of America, Denver, CO.

Busacca, A.J. and E.V. McDonald, 1994. Regional sedimentation of late Quaternary loess on the Columbia Plateau: sediment source areas and loess distribution patterns, in R. Lasmanis and E.S. Cheney, Eds., *Regional Geology of Washington State*, Washington Division of Geology and Earth Resources Bulletin 80, 181–190.

Busacca, A.J. and J.A. Montgomery, 1992. Field-landscape variation in soil physical properties of the Northwest dryland crop production region, in R. Veseth and B. Miller, Eds., *Precision Farming for Profit and Conservation, 10th Inland Northwest Conservation Farming Conference Proceedings*, Washington State University, Pullman, 8–18.

Busacca, A.J., C.A. Cook, and D.J. Mulla, 1993. Comparing landscape-scale estimation of soil erosion in the Palouse using Cs-137 and RUSLE, *J. Soil Water Conserv.*, 48, 361–367.

Busacca, A.J., D.K. McCool, R.I. Papendick, and D.L. Young, 1985. Dynamic impacts of erosion processes on productivity of soils in the Palouse, in *Erosion and Soil Productivity, Proceedings of the National Symposium on Erosion and Soil Productivity*, New Orleans, 10–11 December 1984, American Society of Agricultural Engineers, St. Joseph, MI, 152–169.

Caldwell, H.H., 1961. The Palouse in diverse disciplines, *Northwest Sci.*, 35, 115–121.

Daubenmire, R., 1970. Steppe Vegetation of Washington, Tech. Bull. 62, Washington State Agricultural Experiment Station, Washington State University, Pullman, p. 131.

Donaldson, N.C., 1980. *Soil Survey of Whitman County, Washington*, UDSA-SCS and Washington State University Agricultural Research Center, U.S. Government Printing Office, Washington, D.C.

Elliott, W.J., A.M. Liebenow, J.M. Laflen, and K.D. Kohl, 1989. A Compendium of Soil Erodibility Data from WEPP Cropland Soil Field Erodibility Experiments 1987 & 1988, NSERL Report No. 3, USDA-ARS National Soil Erosion Research Laboratory, W. Lafayette, IN.

Fiez, T.E., W.L. Pan, and B.C. Miller, 1995. Nitrogen use efficiency of winter wheat among landscape positions, *Soil Sci. Soc. Am. J.*, 59, 1666–1671.

Flanagan, D.C. and M.A. Nearing, Eds., 1995. USDA-Water Erosion Prediction Project—Hillslope Profile and Watershed Model Documentation, NSERL Report No. 10, USDA-ARS National Soil Erosion Research Laboratory, W. Lafayette, IN.

Follett, R.F. and B.A. Stewart, Eds., 1985. *Soil Erosion and Crop Productivity*, American Society of Agronomy, Crop Science Society of America, and Soil Science Society of America, Madison, WI.

Formanek, G.E., D.K. McCool, and R.I. Papendick, 1984. Freeze-thaw and consolidation effects on strength of a wet silt loam, *Trans. ASAE*, 26(6), 1749–1752.

Fosberg, M.A., D.J. Walker, M.K. Brodahl, A.L. Falen, and D.J. Mital, 1987. Assessing Wheat Yield Response to Topsoil Depth, University of Idaho Miscellaneous Series No. 99, 485 pp.

Foster, G.R. and L.J. Lane, 1987. User Requirements: USDA-Water Erosion Prediction Project (WEPP), NSERL Report No. 1, USDA-ARS National Soil Erosion Research Laboratory, W. Lafayette, IN, 43 pp.

Frazier, B.E. and A.J. Busacca, 1987. Satellite assessment of erosion, in L.F. Elliott, Ed., *STEEP—Conservation Concepts and Accomplishments*, Washington State University Press, Pullman, 579–584.

Gentry, H.R., 1984. *Soil Survey of Grant County, Washington*, UDSA-SCS and Washington State University Agricultural Research Center, U.S. Government Printing Office. Washington, D.C.

Hoag, D.L., 1984. An Evaluation of USDA Commodity Program Incentives for Erodible Land Retirement, Ph.D. dissertation, Washington State University, Pullman.

Horner, G.M., A.G. McCall, and F.G. Bell, 1944. Investigations in erosion control and the reclamation of eroded land at the Palouse Conservation Experiment Station, Pullman, Washington, 1931–42, U.S. Department of Agriculture, Soil Conservation Service Tech. Bull. 860.

Kaiser, V.G. 1967. Soil erosion and wheat yields in Whitman County, Washington, *Northwest Sci.*, 41, 86–91.

Kaiser, V.G., W.A. Starr, and S.B. Johnson, 1961. Types of topography as related to land use in Whitman County, Washington, *Northwest Sci.*, 25, 69–75.

Kardos, L.T., P.I. Vlasoff, and S.N. Twiss, 1944. Factors contributing to landslides in the Palouse region, *Soil Sci. Soc. Am. Proc.*, 8, 437–439.

Kirkham, V.R.D., M.M. Johnson, and D. Holm, 1931. Origin of the Palouse hills topography, *Science*, 73, 207–209.

Knisel, W.G., Ed., 1980. CREAMS: A Field Scale Model for Chemicals, Runoff, and Erosion from Agricultural Management Systems, U.S. Department of Agriculture, Conservation Research Rep. 26.

Kok, H. and D.K. McCool, 1990. Quantifying freeze-thaw induced variability of soil strength, *Trans. ASAE*, 33(2), 501–506.

Laflen, J.M., G.R. Foster, and C.A. Onstad, 1985. Simulation of individual-storm soil loss for modeling the impact of soil erosion on crop productivity, in S.A. El-Swaify, W.C. Moldenhauer, and A. Lo, Eds., *Soil Erosion and Conservation*, Soil Conservation Society of America, Ankeny, IA, 285–295.

Langdale, G.W. and W.D. Schrader, 1982. Soil erosion effects on soil productivity of cultivated cropland, in B.L. Schmidt, R.R. Allmaras, J.V. Mannering, and R.I. Papendick, Eds., *Determinants of Soil Loss Tolerance*, American Society of Agronomy, Spec. Publ. 45, American Society of Agronomy, Madison, WI, 41–51.

Larney, F.J., R.C. Izaurralde, H.H. Janzen, B.M. Olson, E.D. Solberg, C.W. Lindwall, and M. Nyborg, 1995. Soil erosion-crop productivity relationships for six Alberta soils, *J. Soil Water Conserv.*, 50, 87–91.

Lenfesty, C.D., 1967. *Soil Survey of Adams County, Washington*, UDSA-SCS and Washington Agricultural Experiment Station, U.S. Government Printing Office, Washington, D.C.

Lloyd, C.H. and G.W. Eley, 1952. Graphical solution of probable soil loss formula for Northeastern region, *J. Soil Water Conserv.*, 7, 189–191.

Lotspeich, F.B. and H.W. Smith, 1953. Soils of the Palouse loess: I. The Palouse Catena, *Soil Sci.*, 76, 467–480.

McCool, D.K. and G.O. George, 1983. A second-generation adaptation of the Universal Soil Loss Equation for Pacific Northwest drylands, Paper No. 83-2066, ASAE, St. Joseph, MI.

McCool, D.K., W.H. Wischmeier, and L.C. Johnson, 1974. Adaptation of the Universal Soil Loss Equation to the Pacific Northwest, Paper 74-2523, ASAE, St. Joseph, MI.

McCool, D.K., L.C. Brown, G.R. Foster, C.K. Mutchler, and L.D. Meyer, 1987. Revised slope steepness factor for the Universal Soil Loss Equation, *Trans. ASAE*, 30(5), 1387–1396.

McCool, D.K., G.O. George, M. Freckleton, C.L. Douglas, Jr., and R.I. Papendick, 1993. Topographic effect on erosion from cropland in the Northwestern Wheat Region, *Trans. ASAE*, 36(4), 1067–1071.

McCool, D.K., J.A. Montgomery, A.J. Busacca, and B.E. Frazier, 1997. Soil degradation by tillage movement, Advances in Geo. Ecology, 31, 327–332.

McDaniel, P.A. and A.L. Falen, 1994. Temporal and spatial patterns of episaturation in a Fragixeralf landscape, *Soil Sci. Soc. Am. J.*, 58, 1451–1457.

Montgomery, J.A., 1993. Measurement and Prediction of Long-Term Soil-Landscape Change in the Palouse Region Induced by Tillage and Water Erosion, Ph.D. Dissertation, Washington State University, Pullman, 253 pp.

Montgomery, J.A., A.J. Busacca, B.E. Frazier, and D.K. McCool, 1997. Evaluating soil movement using cesium-137 and the Revised Universal Soil Loss Equation, *Soil Sci. Soc. Am. J.*, 61, 571–579.

Mulla, D.J., 1986. Distribution of slope steepness in the Palouse region of Washington, *Soil Sci. Soc. Am. J.*, 50, 1401–1406.

Musgrave, G.W., 1947. The quantitative evaluation of factors in water erosion: a first approximation, *J. Soil Water Conserv.*, 2, 133–138.

Pan, W.L. and A.G. Hopkins, 1991. Winter barley development, N and P use: evidence of water stress-induced P deficiency in an eroded toposequence, *Plant Soil*, 135, 9–19.

Pan, W., T. Fiez, B. Miller, and A. Kennedy, 1992. Variable soil biological and chemical factors to consider in landscape management, in R. Veseth and B. Miller, Eds., *Precision Farming for Profit and Conservation, 10th Inland Northwest Conservation Farming Conference Proceedings*, Washington State University, Pullman, 20–25.

Papendick, R.I. and R. Veseth, Eds., 1996. Northwest Columbia Plateau Wind Erosion Air Quality Project Interim Report, Misc. Publ. No. MISC0184, Washington State University, College of Agriculture and Home Economics, Pullman.

Papendick, R.I., D.L. Young, D.K. McCool, and H.A. Krauss, 1985. Regional effects of soil erosion on crop productivity—Pacific Northwest, in R.F. Follett and B. A. Stewart, Eds., *Soil Erosion and Crop Productivity*, ASA-CSSA-SSSA, Madison, WI.

Pawson, W.W., 1961. Economics of Cropping Systems and Soil Conservation in the Palouse, Agricultural Experiment Station Bull. No. 2, Washington State University, Pullman.

Pierson, F.B. and D.J. Mulla, 1990. Aggregate stability in the Palouse region of Washington: effect of landscape position, *Soil Sci. Soc. Am. J.*, 54, 1407–1412.

Pikul, J.L., Jr. and R.R. Allmaras, 1986. Physical and chemical properties of a Haploxeroll after fifty years of residue management, *Soil Sci. Soc. Am. J.*, 50, 214–219.

Pimentel, D., C. Harvey, P. Resosudarmo, K. Sinclair, D. Kurz, M. McNair, S. Crist, L. Shpritz, L. Fitton, R. Saffouri, and R. Blair, 1995. Environmental and economic costs of soil erosion and conservation benefits, *Science*, 267, 1117–1123.

Rasmussen, P.E. and W.J. Parton, 1994. Long-term effects of residue management in wheat-fallow. I. Inputs, yield, and soil organic matter, *Soil Sci. Soc. Am. J.*, 58, 523–530.

Renard, K.G., G.R. Foster, G.A. Weesies, D.K. McCool, and D.C. Yoder, Coordinators, 1997. Predicting Soil Erosion by Water: A Guide to Conservation Planning with the Revised Universal Soil Loss Equation (RUSLE), USDA Agriculture Handbook No. 703, 404.

Richardson, C.A., E.V. McDonald, and A.J. Busacca, 1997. Luminescence dating of loess from the Northwest United States, *Quat. Sci. Rev. (Quat. Geochronol.)*, 16, 403–415.

Rickman, R.W., E.L. Klepper, S.E. Waldman, and D.K. McCool, 1994. Predicting green canopy cover of cereal grasses for a RUSLE crop database. *Abstr. 8th Conference of the International Soil Conservation Organization*, New Delhi, India, 86–87.

Rieger, S. and H.W. Smith, 1955. Soils of the Palouse loess: II. Development of the A2 horizon, *Soil Sci.*, 79, 301–319.

Ringe, D., 1970. Sub-loess basalt topography in the Palouse hills, southeastern Washington, *Geol. Soc. Am. Bull.*, 81, 3049–3060.

Rockie, W.A., 1950. Snowdrift erosion in the Palouse, *Northwest Sci.*, 24, 41–42.

Rodman, A.W., 1988. The Effect of Slope Position, Aspect, and Cultivation on Organic Carbon Distribution in the Palouse, Master's thesis, Washington State University, Pullman, 164 pp.

Saxton, K.E., L.D. Stetler, and L.B. Horning, 1996. Predicting and controlling wind erosion and dust emissions, in R.I. Papendick and R. Veseth, Eds., Northwest Columbia Plateau Wind Erosion Air Quality Project Interim Report, Misc. Publ. No. MISC0184, Washington State University, College of Agriculture and Home Economics, Pullman, 7–12.

Schertz, D.L., 1983. The basis for soil loss tolerance, *J. Soil Water Conserv.*, 38, 10–14.

Smith, D.D., 1941. Interpretation of soil conservation data for field use, *Agric. Eng.*, 22, 173–175.

Smith, D.D. and D.M. Whitt, 1948. Evaluating soil losses from field areas, *Agric. Eng.*, 29, 394–398.

Stetler, L.D. and K.E. Saxton, 1996. Wind erosion, PM10 emissions, and dry-land farming on the Columbia Plateau, in N. Lancaster, Ed., *Special Issue: Desert Geomorphology, Earth Surface Processes and Landforms*, 21(7), 673–685.

Taylor, D.B., 1982. Evaluating the Long Run Impacts of Soil Erosion on Crop Yields and Net Farm Income in the Palouse Annual Cropping Region of the Pacific Northwest, Ph.D. dissertation, Washington State University, Pullman.

U.S. Department of Agriculture, 1978. *Palouse Cooperative River Basin Study*, Soil Conservation Service, Forest Service, and Economics, Statistics and Cooperative Service, U.S. Government Printing Office, Washington, D.C., 182 pp.

Van Doren, C.A. and L.J. Bartelli, 1956. A method of forecasting soil losses, *Agric. Eng.*, 37, 335–341.

Van Klaveren, R.W. and D.K. McCool, 1987. Hydraulic Erosion Resistance of Thawing Soil, Paper No. 87-2602, ASAE, St. Joseph, MI.

Van Liew, M.W. and K.E. Saxton, 1983. Slope steepness and incorporated residue effects on rill erosion, *Trans. ASAE*, 26, 1738–1743.

Verity, G.E. and D.W. Anderson, 1990. Soil erosion effects on soil quality and yield, *Can. J. Soil Sci.*, 70, 471–484.

Walker, D.J., D.L. Young, M.A. Fosberg, A.J. Busacca, K.E. Saxton, D.L. Carter, and B.E. Frazier, 1987. Long term productivity benefits of soil conservation, in L.F. Elliott, Ed., *STEEP—Conservation Concepts and Accomplishments*, Washington State University Press, Pullman, 9–39.

Wetter, F., 1977. The Influence of Topsoil Depth on Yields, Tech. Note AGRON-10, Soil Conservation Service, Colfax, WA.

Williams, J.D., D.E. Wilkins, D.K. McCool, L.L. Baarstad, E.L. Klepper, and R.I. Papendick, 1998. A new rainfall simulator for the Pacific Northwest, *Applied Engineering in Agriculture*, ASAE, 14(3), 243–247.

Wischmeier, W.H., 1973. Conservation tillage to control water erosion, in *Conservation Tillage, Proceedings of the National Conservation Tillage Conference*, Des Moines, IA, Soil Conservation Society of America, 133–141.

Wischmeier, W.H. and D.D. Smith, 1965. Predicting Rainfall-Erosion Losses from Cropland East of the Rocky Mountains: Guide For Selection of Practices for Soil and Water Conservation, U.S. Department of Agriculture, Agricultural Handbook 282.

Wischmeier, W.H. and D.D. Smith, 1978. Predicting Rainfall Erosion Losses: A Guide to Conservation Planning, U.S. Department of Agriculture, Agricultural Handbook 537.

Wischmeier, W.H., C.B. Johnson, and B.V. Cross, 1971. A soil erodibility nomograph for farmland and construction sites, *J. Soil Water Conserv.*, 26, 189–193.

Young, D.L., D.B. Taylor, and R.I. Papendick, 1985. Separating erosion and technology impacts on winter wheat yields in the Palouse: a statistical approach, in *Erosion and Soil Productivity, Proceedings of the National Symposium on Erosion and Soil Productivity*, New Orleans, 10–11 December 1984, American Society of Agricultural Engineers, St. Joseph, MI, 130–142.

Zingg, A.W., 1940. Degree and length of land slope as it affects soil loss in runoff, *Agric. Eng.*, 21, 59–64.

4 Residue Management Issues for Conservation Tillage Systems

Lloyd F. Elliott, Diane E. Stott, Clyde L. Douglas,
Robert I. Papendick, Gaylon S. Campbell,
and Hal Collins

CONTENTS

ABSTRACT—Management practices that will influence crop residue persistence and residue efficacy for soil erosion protection include no-till vs. reduced tillage, fall seeding vs. spring seeding, prediction of residue longevity and groundcover, and utilization of approaches such as low-input composting. These practices also offer the potential for improving soil quality. No-till seeding alleviates wind and water erosion in many cases. However, when crop residues are limiting, minimum tillage seeding that utilizes cloddiness for erosion protection may be more effective. Continuous spring seeding in lieu of fall seeding is an excellent method for combating wind and water erosion and for weed control. RESMAN is a model that was developed for predicting residue persistence for erosion control. It is used widely for this purpose. Where heavy crop residues discourage successful conservation cropping systems, incorporation of an approach such as low-input, on-farm composting could be a valuable tool.

4.1   INTRODUCTION

Reducing soil erosion by water or wind is a national high-priority issue. Crop residue management, amount, and quality will play an important role in controlling soil erosion (Stott et al., 1995). Buried residues offer less soil physical protection and will decompose more rapidly than residues on the soil surface. Surface residues will reduce wind and water energy, resulting in less soil movement when compared with similar unprotected soil. Based on these principles, cropping practices that leave the largest amount of residues on the soil surface generally will provide the most protection against soil erosion.

Another national high-priority issue is maintaining or increasing soil quality. Soil quality has been defined as the capacity of soil to produce healthy and nutritious crops, resist erosion, and reduce the impact of environmental stresses on plants (Papendick and Parr, 1992; Elliott and Lynch, 1994). The management of crop residues for conservation tillage systems will play a major role in maintaining or increasing soil quality, with no-till cropping systems being most effective. Crop residues and roots are key contributors to soil organic matter content (Smith and Elliott, 1990). A residue and soil management cropping system that uses reduced tillage when compared with a previous more intensive tillage will improve soil quality because tillage decreases the soil organic matter content. The great difficulty has been to identify measurements that are useful for following temporal changes in soil quality. Several measurements have been proposed (Papendick, 1991; Doran et al., 1994; Lynch and Elliott, 1997). Unfortunately, for many of these measurements to show differences, several years of management are required to be significant. Enzyme and enzyme product measurement and gene presence and gene regulation show promise for providing temporal measurements in soil quality (Lynch and Elliott, 1997).

The cropping system will govern the effectivity of conservation tillage systems for erosion control and ultimately the effect on soil quality. The cropping system, degree of tillage and residue, and residue management will largely influence the degree of erosion control. The factors influencing these choices include fall seeding vs. spring seeding, no-till vs. reduced tillage, prediction of residue longevity and groundcover, and the possible utilization of alternate residue management approaches such as the development of low-input, on-farm composting technology of high C/N ratio crop residue substrates and placing the compost back on the land. The possibilities for these choices are discussed in the following text.

4.2   THE CASE FOR FALL SEEDING AND SPRING SEEDING AND TILLAGE EFFECTS

In the Pacific Northwest, residue cover for erosion control is generally most critical when the fall crop is planted because of the high potential for soil erosion by water during the winter and wind erosion in the fall. Water and wind erosion rates in a given precipitation zone are dictated largely by management history that dates back to other times and intervals in the rotation or crop sequence, including at the time of planting. The main concerns with residue management are inadequate residues for fall planting following low-residue-producing crops in the high precipitation zone and excessive winter wheat residue during seedbed preparation in an annual cropping sequence.

4.2.1   WINTER WHEAT FOLLOWING A LOW-RESIDUE CROP

Low amounts of residues are common following pea and lentil crops and, in some cases, after spring barley or wheat. Quantities can be especially limited in areas of the landscape that have low fertility, e.g., the mid- and upper-slopes and the ridge tops where the erosion potential is high. Pea, lentil, and, to a lesser extent, barley straw also decompose more readily than wheat straw (Stott et al., 1995), which reduces the amount of residue at fall planting time.

Through research, two residue management approaches have evolved that are generally effective for erosion control in fall wheat following low-residue conditions. In one, a no-till drill is used that simultaneously seeds and deep-bands fertilizer with little disturbance to the soil or residue. The second approach is a minimum tillage system that begins with a shank fertilizer operation followed by sowing with a conventional drill.

Drill requirements with the no-till system include ability to operate on hillsides, penetrate hard, dry soil, and cut through residues to avoid "hair pinning" of residues in the seed row or plugging the openers. There are a variety of one-pass drills that can meet these criteria and be affordable to many producers. One concern with no-till seeding is the lack of surface roughness; further, in some situations relying on residues alone may be insufficient for adequate control of erosion. A trade-off would be a drill that produces some roughness in dry soil even at the cost of burying some residue.

The no-till system can be facilitated by uniform distribution of residues behind the combine which may require special attention during the harvest operation or prior to planting. Combine straw-spreading attachments or modifications can improve residue distribution and these are generally available for most equipment. Also, straw can be spread more evenly by a harrowing operation after harvest perpendicular or at an angle to the direction of the combine windrow.

A minimum tillage system that combines cloddiness and residues for erosion control is gaining acceptance by many farmers for planting fall wheat in low-residue conditions. The direct shank-and-seed system can realistically achieve residue levels of 20 to 30% cover following pea and lentil crops. The practice is observed to be more effective for erosion control than that with a double-disk, no-till drill, which leaves more residue on the surface. A heavy-duty shank fertilizer applicator is used to deep-band fertilizer without any prior tillage. Growers apply herbicide for weed control and then seed with their conventional drills. There are a number of types of direct-shank fertilizer applicators available either from chemical dealers or versions of chisel plows or cultivators developed independently by growers. A one-pass version of the shank-and-seed approach known as a chisel drill has also been developed. Fertilizer is placed with chisel shanks, and the crop is seeded with an attached double-disk drill in one operation. Herbicides are applied before fertilizing and seeding.

4.2.2 FALLOW

Cereals are the main crop in the fallow zones. Residue amounts are lower initially and decomposition losses are greater than in annual cropping systems. The lower amounts of residue in a fallow system can limit surface cover at planting time. With conventional farming methods, using a primary tillage in the fall or early spring followed by three to five summer rod-weedings to control weeds often results in very little residue cover left after the fall wheat is planted. Special efforts are needed to achieve adequate cover for wind and water erosion control. Approaches for residue management include chemical weed control as a substitute for tillage and modifications of tillage practices.

Farmer observations indicate that fall tillage accelerates residue disappearance during fallow. Delaying tillage as late as possible in the spring tends to preserve residues longer. Fall tillage is used to kill Russian thistle (*Salsola iberica*), which otherwise grows to maturity in fields after wheat harvest. Fall chiseling is also used in many areas to promote water infiltration into frozen soils.

Russian thistle has a deep taproot and depletes water in the fall in the lower parts of the profile, which is recharged more slowly and only in wetter years. However, in many fields weed populations are often sparse with plants spaced a few to 10 m apart. New sensing technology may allow these weeds to be spot sprayed, which would be more economical than tillage and would eliminate the need for fall tillage.

Fall tillage for frozen soil infiltration can be modified by widening the shank spacings from the conventional 12 to 24 in. to 60 in. or more and keeping the tillage on the slope contour as much as possible. These wider spacings will disturb the residues less and promote their longevity on the surface. Spring tillage can be delayed by application of a nonselective herbicide to kill early spring growth of weeds while leaving stubble intact. The primary tillage can be initiated when second weed growth begins and to set the tillage depth for the fallow. Secondary tillage operations, such as rodweeding, should be limited to those essential-to-kill weeds. Efforts should be made to minimize or combine tillage operations, such as cultivators and rod-weeders, heavy-duty shank fertilizer applicators, and chisel choppers. All of these practices, contrasted to conventional approaches, tend to increase residue levels at fall planting time and help farmers meet their conservation plan requirements.

Chemical fallow without tillage has not gained much acceptance with winter wheat producers in the dryland areas. The cost of chemicals for the required two to three applications is prohibitive to many growers. No-till fallow reduces seed zone water at planting time compared with tillage fallow which insulates the soil and prevents capillary water flow to the surface. Lower seed zone moisture delays fall wheat establishment and can reduce wheat yields by 40% compared with early established stands. Chemical fallow should, however, not be ruled out as an option in highly erodible fields or parts of the landscape that are highly susceptible to wind erosion. No-till culture is probably the best opportunity for stabilizing these soils and restoring their fertility and quality for meeting long-term productivity and environmental goals.

4.2.3 HANDLING EXCESSIVE RESIDUES WITH ANNUAL CROPPING

Excessive residues following a winter wheat crop can be difficult to manage for annual cropping. They can interfere with fall planting in the case of recropping and with spring planting and weed control. In the intermediate precipitation zone, heavy residues can interfere with fallow operations. Management opportunities include residue reduction techniques such as chopping, harrowing, light disking, or shallow incorporation. Once residues are broken down in size, they are easier to handle during subsequent operations.

Tillage tools such as disks or angled blade chisels when operated at the proper depth incorporate 30 to 40% of the residues in the shallow layers. The residue retains some effectiveness for erosion control and interferes less with subsequent tillage and planting operations. Such implements usually feature a disk assembly mounted in front of a chisel plow, each with independent hydraulic controls. Adjustable disk gangs cut through the crop residue ahead of chisel shanks, which penetrate more deeply and loosen the soil. On some machines, a second gang of angled, adjustable disks follows in the rear to smooth the ridges and evenly distribute the remaining surface residues.

4.2.4 CONTINUOUS SPRING CROPPING

More recently, a few growers in the low-precipitation zone have implemented continuous, no-till spring wheat cropping systems. The crop is planted usually as early as possible in late winter, which can be from mid-February to mid-March when the weather allows and the soil is firm. A no-till drill is used that seeds and deep-bands fertilizer in a one-pass operation. In one documented case, this practice has been applied by a grower on the same field for 10 years in a 9-in. annual precipitation zone. The no-till practice keeps cover on the soil at all times of the year and virtually eliminates wind and water erosion on a sandy soil. Weeds are controlled with two chemical applications that are well within the economics of this production system. Hard red spring wheat yields have averaged 20 bu/acre/year with high protein content compared with 35 bu/acre every other year with the winter wheat/fallow system. Winter annual grass weeds, such as downy brome and jointed goatgrass, that are a major problem with the winter wheat system are virtually eliminated with the continuous spring cropping. This practice merits greater attention in the dryland areas that are prone to serious wind erosion. A key to expanding continuous spring wheat cropping is the development of varieties with improved resistance to Hessian fly and cold tolerance for establishment and growth at low temperatures in the early spring.

4.3   CROP RESIDUE DECOMPOSITION

4.3.1   FACTORS AFFECTING CROP RESIDUE DECOMPOSITION

Soil protection by crop residues occurs by two processes. First, the residues absorb water and wind energy and, second, after a period of decomposition of crop residues, the bodies of the microbes and metabolic by-products bind soil particles, increasing soil resistance to erosion and improving water infiltration. However, legume residues do not protect so well because they decompose rapidly and may not protect during critical periods. The need for surface-managed crop residues for the prevention of soil erosion was obvious at the initiation of STEEP. Moreover, it was apparent that efforts were needed to predict crop residue decomposition at the soil surface, to determine microbial effects on plant growth, to study mechanisms controlling soil erodibility, and as an aid to provide crop residue and soil management approaches during critical periods.

The development of a predictive effort for crop residue decomposition in the field that would be functional across climatic zones required measuring the factors that control crop residue decomposition. There are a large number of inputs required for mechanistically predicting crop residue decomposition, and many are interdependent. For example, residues left on the soil surface will affect both soil temperature and moisture (Parr and Papendick, 1978). Residue placement and handling will result in quite different environmental effects and in residue decomposition rates. Surface residues disappear at about two thirds the rate of buried residues (Brown and Dickey, 1970; Douglas et al., 1980). Partially or completely incorporated residues have more intimate soil contact and are not subject to the temperature extremes of surface residues. Also, the apparent rate of residue decomposition is affected greatly by the method of measurement. Caution has to be exercised with data collected by bags because the mesh bag method underestimates residue decomposition by about 30% when compared with residue placed directly in or on the soil (Witkamp and Olson, 1963; Wiegert and Evans, 1964; Wieder and Long, 1982). Residue decomposition was affected little by placement if water and temperature were held constant (Stott et al., 1986). Jawson and Elliott (1986) showed that soil inoculum was not needed to provide organisms for straw decomposition. The residues contained sufficient indigenous flora for this purpose.

The chemical composition of the crop residue is another important regulator of decomposition (Summerall and Burgess, 1989). This has been known for some time when comparing, for example, a cereal such as wheat straw with a legume such as pea straw. However, there are large differences in the decomposability of wheat straws and components of wheat straw (Reinertsen et al., 1984; Collins et al., 1990a). The individual components of the residue affect decomposition rate. The work by Collins et al. (1990b) with "Daws" winter wheat showed that the ratio of component plant parts (leaf, leaf sheath, chaff, and stem) remained relatively constant between growing seasons and yield differences. The average ratio of the leaf, leaf sheath, chaff, and stem components was 14:20:28:38. The total C, total N, soluble C, and nonstructural carbohydrate concentrations of the wheat residues were not correlated with residue or grain yields (Collins et al., 1990b). The decomposition rates of the individual parts were closely related to the carbohydrate, lignin, C, and N contents. This relationship did not hold true when the residue parts were mixed together prior to incubation and decomposition. Residue mixes decomposed about 25% more rapidly than what was predicted from the decomposition rates of the individual parts (Collins et al., 1990b).

The potential decomposition rate of wheat straw can be based on the size of the readily available C and N pools (Knapp et al., 1983a, b; Reinertsen et al., 1984). From this work, it was postulated that microbial extracellular materials such as polysaccharides might dominate the aggregation process shown by decomposing straw if the wheat straw contained low N, and if alternate sources of N were unavailable. Elliott and Lynch (1984) aerobically degraded three wheat straws containing 1.09, 0.5, and 0.25% N in the absence of added N. The 0.25% N straw treatment produced significantly more aggregation in the soils tested than the other treatments. The 0.5% N straw treatment generally caused more aggregation than the 1.09% N straw. The largest microbial biomass would be generated from the straw containing the most N; thus, these results confirm the postulate that the increased aggregation resulted from the microbial production of extracellular gums by the smaller microbial biomass. Electron micrographs also showed more gum production on the 0.25% N-containing straw. Polysaccharides have been implicated in the soil aggregation process (Tisdall and Oades, 1982). The results of these studies show that there is potential for improving soil aggregation through residue management on the soil surface and by reducing tillage, because tillage increases the rate of soil organic matter mineralization (Rovira and Greacen, 1957). Improved soil aggregation increases water infiltration, resistance to wind and water erosion, and probably soil productivity.

Hunt (1978) says grassland substrates cannot be treated as homogeneous, nor can each of its components be treated separately. He suggests they be considered as heterogeneous materials that consist of two fractions. Collins, et al. (1990a) and Douglas and Rickman (1992) working with cereals, suggest the same. They, along with Berg and Agren (1984), Berg (1986), and Janzen and Kucey (1988), suggest there is a labile or rapidly decomposing fraction (proteins, sugars, starches) and a resistant or slowly decomposing fraction (celluloses, lignins, fats, tannins, waxes). Others, such as van Veen and Paul (1981), separated decomposable substrates into three fractions, each progressively more resistant to decomposition; carbohydrates and proteins, cellulose and hemicellulose, and lignin. Reinertsen et al. (1984) and Stroo et al. (1989) also considered cereal substrates to consist of three separate fractions, carbon (C) pools, based on availability to microorganisms. They designated the pools as readily available, intermediately available (cellulose, hemicellulose), and resistant (lignin). With their model, Stroo et al. (1989) were able to predict wheat residue decomposition across climatic zones.

The amount of readily available C and N controls the size of the initial microbial biomass and the initial decomposition rate. Readily available C and N content of the residues increases as the C/N ratio decreases (Reinertsen et al., 1984). Residues high in total N tend to be high in soluble N (Iritani and Arnold, 1960). Increasing the N concentration of a substrate results in increasing the decomposition rate of the easily soluble materials (Douglas and Rickman, 1992).

Most models treat decomposers entirely as microbes. When fauna was excluded from litter bags, decomposition rates in grasslands, in contrast to forests, are little affected (Curry, 1969). Stroo et al. (1989) showed fauna accounted for 5% or less of the CO_2 respired. Distinguishing each group of decomposers and their relative contributions would be almost impossible with current technology (bacteria, fungi, actinomycetes, fauna, etc.). This precision is unnecessary for current predictive needs. Models generally assume that under optimal environmental conditions there is a microbial population present to sustain the maximum decomposition rate.

There must be favorable moisture and temperature conditions for activity. However, each group, and even subgroup, of decomposer has an optimum and range of climatic conditions in which it is active. Fluctuating climatic conditions, especially those that go beyond the range of the decomposer, are more deleterious to decomposers than constant conditions (Parr and Papendick, 1978). For practical considerations in the field, it is unlikely much biological decomposition occurs above 20°C because above this temperature, in the field, residues are usually too dry to decompose. Effects of climatic conditions are different even with different phases of decomposition. For example, Stott et al. (1986) found that low water potentials or low temperatures had significant effects on microbial activity only during the initial phase of decomposition. Thus, both long-term and diurnal fluctuations of temperature and moisture must be inputs.

4.4   MODELING CROP RESIDUE DECOMPOSITION

Generally, decomposition is evaluated across some time frame. Decomposition can be related to degree days (DD), calculated from air temperature (Douglas and Rickman, 1992). Degree days are determined by measuring the daily mean air temperature in degrees centigrade and summing over the desired period. If the mean daily temperature is less than 0°C, it is considered as zero. Zero is used as a base value because of reports by Wiant (1967) that microbial reactions follow the Van't Hoff and Arrhenius laws at temperatures below 40°C. Reiners (1968) confirms that this is true down to 0°C. However, these relationships must be viewed with caution because dramatic changes occur in the flora makeup and activity as temperatures change. Stott et al. (1986) showed the response to temperature was likely related to changes in the microflora as temperature was changed and this was the reason the system did not respond according to Van't Hoff's law.

Residue decomposition when evaluated in relation to time is quite different across the U.S. because of large variations in temperature and moisture (Stott et al., 1990). However, a DD is a DD regardless of where one is in the world. Decomposition can still be related back to time with knowledge of the time–DD relationship for the selected area, which is very easy to determine.

The decomposition model by Douglas and Rickman (1992), based on DD, can be used anywhere in the U.S. where air temperature is determined. All that is required is air temperature (temperature factor), initial nitrogen (N) content of the decomposing residue, residue placement (surface or buried), and if the residue is decomposing under a crop or in fallow (a moisture factor). This model simulates the initial (rapid rate, easily decomposable materials) decomposition rate depending on the N content of the decomposing residue. For example, if the N content is less than 0.55%, initial decomposition occurs at a slower rate than if the N content is greater than 0.55%. This initial (rapid) decomposition rate occurs in approximately 1000 DD, which corresponds roughly to September through November in the Pendleton, Oregon area. This model has been verified by comparing output to data from Arkansas, Indiana, Texas, Washington, Idaho, Missouri, and two places in Canada for wheat and for red clover in Missouri. It should be pointed out that the model is empirical and that it will be affected by changes in moisture and by composition of the substrate. The availability of N will vary with substrate, for example.

4.5   RESIDUE DECOMPOSITION MODELS IN EXPERT SYSTEMS AND EROSION MODELS

In 1989, Stroo et al. published a mechanistic-based model for wheat residue decay. The model simulates decay under constant environmental conditions using C and N dynamics. It then determines the impact of environmental conditions, calculating the fraction of an optimum "decomposition day" occurring in a 4-h period. Information concerning initial residue C and N pool availability is required for this model.

Once the theoretical model for crop residue decomposition was developed, the next step was to simplify the model for use as a component in an expert system on residue management (RESMAN) (Stott et al., 1988; 1995; Stott and Rogers, 1990; Stott, 1991). The goal in developing RESMAN was to incorporate residue decomposition knowledge with site- and situation-specific tillage considerations including residue burial. Inputs for the expert system needed to be relatively simple and readily available to a wide variety of users. Another feature was a user-specified need for quick runtimes; thus, the expert system was unable to deal with the complexity of a true research model.

RESMAN, since its release in 1990, has been used widely by industry personnel, extension and soil conservation advisors, and private consultants throughout the U.S., Canada, and several other countries to develop crop residue management strategies for soil conservation. Due to its wide acceptance, the theory and equations used in RESMAN were incorporated into several erosion models being developed by the U.S. Department of Agriculture (USDA), including: RUSLE (revised universal soil loss equation), RWEQ (revised wind erosion equation), WEPP (water erosion prediction project), and WEPS (wind erosion prediction system). RUSLE was implemented by the USDA-Natural Resources Conservation Service (NRCS) in the summer of 1995. RWEQ was to be implemented a year later. WEPP and WEPS, utilizing new erosion prediction technologies, are expected to be completed and implemented by the end of the decade.

4.5.1   THEORY USED IN THE RESMAN

To simulate the decomposition process, the "decomposition day" concept as presented by Stroo et al. (1989) for winter wheat residue decomposition was used as a basis for the residue mass loss calculation. The Stroo et al. (1989) model simulates residue decay under constant environmental conditions using C and N dynamics based on Knapp et al. (1983a, b) and Bristow et al. (1986). The residue C is split into three pools based on availability for use by the soil microbial population and chemically defined. This information is not readily available for a wide variety of crops; thus, in the RESMAN model, the equations describing the C and N dynamics were replaced with a single equation (Stott et al., 1995):

$$M_t = M_y * e^{-\left(R_{opt} * EF\right)}$$ (4.1)

where M_t is the residue mass per unit area remaining on the surface today and M_y is the mass per unit area left on the ground the previous day, R_{opt} is a decomposition constant for a given residue type for the amount of mass lost in 1 day under optimum conditions for activity, and EF is the environmental factor determining the fraction microbial of a decomposition day that has occurred during day t. The value for R_{opt} for a given crop can be calculated from the more mechanistic Stroo et al. (1989) model if the nutrient data is available. Alternatively, R_{opt} can be estimated from field or laboratory studies measuring rates of residue mass loss.

In the field, residue decomposition rates are controlled by environmental factors (Martin and Haider, 1986). Especially important are the water content and temperature (Parr and Papendick, 1978). The effects of water content and temperature on the rate of residue decomposition were assumed to be independent of one another (Stott et al., 1986; Stroo et al., 1989). To estimate the influence of these factors on residue decomposition in the field, the following relationship was used (Stott et al., 1995):

$$EF = Minimum\ (WFC, TFC)$$ (4.2)

where EF is the environmental factor used in Equation 4.1, and WFC and TFC are water and temperature factors, respectively, with normalized values between 0 and 1.

The changes in the water content and temperature within the soil and residue layers are calculated with a simplified version of the Bristow et al. (1986) model. Bristow et al. (1986) divided the residue into five layers and the soil into three layers. Temperature and moisture fluctuations were calculated for each layer at 4-h time intervals. In the RESMAN model, three residue and five soil layers are used, and calculations are done at 24-h time intervals. The data needed to calculate the temperature and water content for the soil and residue layers include daily maximum and minimum air temperatures and precipitation. It is assumed that most of the residue decomposition occurs in the bottom residue layer that interfaces with the soil surface. Thus, WFC and TFC are calculated using the temperature and water content for the bottom residue layer. RUSLE, RWEQ, and WEPP do not calculate the temperature or moisture within the residue layer; therefore, WFC is estimated based on the water content in the top soil layer and TFC is estimated from the air temperature.

The water function (WFC) is calculated as (Linn and Doran, 1984):

$$WFC = \frac{\Theta}{\Theta_{opt}} \quad if\ \Theta_- \leq \Theta_{-opt} \quad or \quad WFC = \frac{\Theta_{-opt}}{\Theta} \quad if\ \Theta_- > \Theta_{-opt} \quad (4.3)$$

where Θ is the actual water content (g water per kilogram) of the residue or soil and Θ_{opt} is the optimum water content. The latter value was set at 3500 g water per kilogram of residue as determined by Stroo et al. (1989) from data published by Stott et al. (1986). This residue water content is equivalent to a water potential of −33 kPa, which is considered optimal for microbial activity (Sommers et al., 1981). RESMAN assumed that the soil texture was silt loam. For that soil textural class, a −33 kPa water potential is equivalent to about 60% water-holding capacity. Assuming a bulk density of 1.2 and 50% porosity, Θ_{opt} for the soil layers was set at 833 g water per kilogram of soil. The submodel used in WEPP allows for variations in soil texture and water absorptive capacities of the soil. This includes changes in the amount of pore space within the soil as compaction and tillage occur.

The temperature function (TFC) is based on an equation for photosynthetic activity (Taylor and Sexton, 1972):

$$\text{TFC} = \frac{2(T+A)^2(T_m+A)^2-(T+A)^4}{(T_m+A)^4} \qquad (4.4)$$

where T is the average temperature in °C, T_m is the optimum temperature, and A is an experimentally derived constant. T is calculated as the mean of the daily minimum and maximum temperatures. If T is plotted against TFC, T_m is the temperature at which TFC equals 1. TFC equals 0 at two points, the first where the average daily temperature is too low for microbial activity and the second where it is too high for activity. The constant, A, is equal to the absolute value of the lower of the two TFC zero values. Stroo et al. (1989) used $T_m = 33$°C, and $A = 6.1$°C for calculating TFC at 4-h time intervals. RESMAN uses average daily temperature for T, rather than 4-h averages; thus, a T_m of 30°C and an A of 0°C were used, based on laboratory data from Stott et al. (1986) and personal unpublished field data. Since Equation 4.4 is a quartic function, TFC was set to 0 when $T < 0.0$ or $T > 42.4$.

Each operation, or pass through a field with a tillage implement, inverts residue under the soil surface, reducing erosion protection. Modeling the effects on percent residue cover of all possible tillage procedures is difficult, since each tillage pass not only turns some of the residue under, but also can, in a few cases, return some of the previously buried residue to the surface. Depth and speed of the equipment operation impacts the amount of residue buried, as does the amount of residue present at the time of tillage. For a specific piece of equipment, the shallower the operating depth, the greater the amount of residue left on the surface, whereas deeper operating depths will bury more residue. Additionally, slower operating speeds tend to leave more residue on the surface. In RESMAN (Version 2.0) and the erosion models, the tillage burial coefficients used were derived from the USDA-SCS-EMI tillage implement list (1992).

The effectiveness of a residue layer in protecting the soil against water erosion is dependent on the percentage of the soil surface covered. Because residue decomposition is calculated in terms of residue mass per unit area, a conversion from residue mass to percent cover is needed. The equation used in RESMAN and the erosion models for this conversion is based on one published by Gregory (1982):

$$C_t = 100 * \left(1 - e^{-m_t * K}\right) \tag{5}$$

where the percent surface area covered (C_t) is a function of the M_t from Equation 4.1 and a constant (K) that is dependent on crop type and represents the area covered by a specific mass of residue.

4.6   LOW-INPUT, ON-FARM COMPOSTING

Crop yields often suffer when conservation tillage systems are implemented (McCalla and Army, 1961; Papendick and Miller, 1977). These yield reductions have been attributed to a variety of problems. These include short-chain fatty acids produced during residue decomposition (Cochran et al., 1977; Lynch et al., 1981); infection by plant pathogens such as *Pythium* sp. (Cook et al., 1980); and colonization of roots by deleterious rhizobacteria (Suslow and Schroth, 1982; Fredrickson and Elliott, 1985; Alstrom, 1987; Schippers et al., 1987). Hair pinning of the residues around the seed can result in poor seedling growth because of poor seed zone environmental conditions (Elliott et al., 1984). Many of these problems are more severe as residue production become heavier. The solution has been to burn the residues. Burning of residues is causing increased public concern because of air pollution. Various options, such as using the residues for heat or power production, have been explored but are not economically feasible. The management of heavy residues for conservation tillage systems has been an unyielding problem in many cases.

However, answers for successful conservation tillage cropping system development are essential for addressing the soil quality issue. Tillage decreases soil organic matter content (Smith and Elliott, 1990) and reduces the size of the microbial biomass (Hassink et al., 1991; Brussaard, 1994). Reduction of these soil properties likely reflect a loss in soil quality (Doran et al., 1994). Tillage hastens the mineralization of root biomass which decreases the amount of readily available C over the longer term for maintenance of microbial biomass and nutrient-cycling capability. Also, use of crop residues in the cropping system is important for maintaining soil quality rather than burning them or using them for generation of heat or some other farm use.

The development of low-input, on-farm composting of high-C/N-ratio residues is providing an innovative residue management option. Composting of crop residues will overcome the negative aspects of using conservation tillage with heavy crop residues and may provide added benefits. The potential benefits of crop residues to soil structure have been demonstrated (Elliott and Lynch, 1984). The possible benefits of compost for alleviating some soilborne diseases has also been shown (Hoitink et al., 1991). These possibilities have not been explored thoroughly. The benefits of compost applications also include fertilizer content, soil conditioning value, and benefits to soil quality (Bangar et al., 1989; Nelson and Craft, 1992; Thomsen, 1993; Zaccheo et al., 1993).

Optimum use of crop residues for conservation tillage systems was not feasible until the development of low-input, on-farm composting of high-C/N-ratio crop residues. Churchill et al. (1995) developed the composting approach for grass seed straw and the system should work for crop residues such as wheat straw also. Studies by Horwath et al. (1995) have explained the mechanism of the process which appears to be rapid delignification. Previously, it was thought that successful composting required a combined substrate C/N ratio of 30/1 or less (Biddlestone et al., 1987).

The low-input, on-farm composting method consists of gathering the straw into large piles at the side of the field. When rainfall occurs, the stacks are turned to allow maximum water intake. The turning, with a front-end loader, also compacts the straw which helps the stack to retain heat. Over winter and early spring the straw is turned when temperatures cool. The compost is ready for field application after about three turns in as little as 16 weeks time (Churchill et al., 1995). Studies on the economics of the process and the value of the compost additions to succeeding crops are not complete.

Low-input, on-farm composting will allow no-till seeding or shallow conservation tillage on fields that have contained heavy residues. These cropping techniques will be possible without the risk of severe yield reductions of the succeeding crop. It is also likely that weed control will be expedited in the absence of the heavy residues. Finally, the lack of extensive tillage and return of the residues should enhance soil quality over systems using conventional practices.

REFERENCES

Alstrom, S., 1987. Factors associated with detrimental effects of rhizobacteria on plant growth, *Plant Soil*, 102, 3–9.

Bangar, K. C., S. Shanker, K. K. Kapoor, K. Kukreja, and M. M. Mishra, 1989. Preparation of nitrogen and phosphorus-enriched paddy straw compost and its effect on yield and nutrient uptake by wheat (*Triticum aestivum* L.), *Biol. Fertil. Soils*, 8, 339–342.

Berg, B., 1986. Nutrient release from litter and humus in coniferous forest soils—a mini review, *Scand. J. Res.*, 1, 359–369.

Berg, B. and G. I. Agren, 1984. Decomposition of needle litter and its organic chemical components: theory and field experiments. III. Long term decomposition in a Scots pine forest, *Can. J. Bot.*, 62, 2880–2888.

Biddlestone, A. J., K. R. Gray, and C. A. Day, 1987. Composting and straw decomposition, in *Environmental Biotechnology*, C. F. Forster and D. A. Wase, Eds., John Wiley & Sons, New York, 135–175.

Bristow, K. L., G. S. Campbell, R. I. Papendick, and L. F. Elliott, 1986. Simulation of heat and moisture transfer through a surface residue–soil system, *Agric. For. Meterol.*, 36, 193–214.

Brown, P. L. and D. D. Dickey, 1970. Losses of wheat straw residue under simulated field conditions, *Soil Sci. Soc. Am. J.*, 34, 118–121.

Brussaard, L., 1994. The ecology of soil organisms in reduced input farming, in *Soil Biota: Management in Sustainable Farming Systems*, C. E. Pankhurst, B. M. Doube, V. V. S. R. Gupta, and P. R. Grace, Eds., CSIRO Information Services, P.O. Box 89, East Melbourne, Victoria 3002, 197–203.

Churchill, D. B., D. M. Bilsland, and L. F. Elliott, 1995. Method for composting grass seed straw residue, *Appl. Eng. Agric.*, 11, 275–279.

Cochran, V. L., L. F. Elliott, and R. I. Papendick, 1977. The production of phytotoxins from surface crop residues, *Soil Sci. Soc. Am. J.*, 41, 903–908.

Collins, H. P., L. F. Elliott, and R. I. Papendick, 1990a. Wheat straw decomposition and changes in decomposability during field exposure, *Soil Sci. Soc. Am. J.*, 54, 1013–1016.

Collins, H. P., L. F. Elliott, R. W. Rickman, D. F. Bezdicek, and R. I. Papendick, 1990b. Decomposition and interactions among wheat residue components, *Soil Sci. Soc. Am. J.*, 54, 780–785.

Cook, R. J., J. W. Sitton, and J. T. Waldher, 1980. Evidence for *Pythium* as a pathogen of direct drilled wheat in the Pacific Northwest, *Plant Dis.*, 64, 102–103.

Curry, J. O., 1969. The decomposition of organic matter in soil. Part I. The role of the fauna in decaying grassland herbage, *Soil Biol. Biochem.*, 1, 235–258.

Doran, J. W., M. Sarrantonio, and R. Janke, 1994. Strategies to promote soil quality and health, in *Management in Sustainable Farming Systems*, C. D. Pankhurst, B. M. Doube, V. V. S. R. Gupta, and P. R. Grace, Eds., Soil Biota, CSIRO, East Melbourne, Victoria, 3002, Australia, 230–247.

Douglas, C. L., Jr. and R. W. Rickman, 1992. Estimating crop residue decomposition from air temperature, initial nitrogen content, and residue placement, *Soil Sci. Soc. Am. J.*, 56, 272–278.

Douglas, C. L., Jr., R. R. Allmaras, P. E. Rasmussen, R. E. Ramig, and N. C. Roager, Jr., 1980. Wheat straw composition and placement effects on decomposition in dry land agriculture of the Pacific Northwest, *Soil Sci. Soc. Am. J.*, 44, 833–837.

Elliott, L. F. and J. M. Lynch, 1984. The effect of available carbon and nitrogen in straw on soil and ash aggregation and acetic acid production, *Plant Soil*, 78, 335–343.

Elliott, L. F. and J. M. Lynch, 1994. Biodiversity and soil resilience, in *Soil Resilience and Sustainable Land Use*, D. J. Greenland and I. Szabolcs, Eds., CAB International, Wallingford, U.K., 353–364.

Elliott, L. F., R. I. Papendick, and V. L. Cochran, 1984. Phytotoxicity and microbial effects on cereal growth, in *Proc. Sixth Manitoba-North Dakota Zero Tillage Workshop*, Bismarck, ND, January 1984, 40–47.

Fredrickson, J. K. and L. F. Elliott, 1985. Effects on winter wheat seedling growth by toxin producing rhizobacteria, *Plant Soil*, 83, 399–409.

Gregory, J. M., 1982. Soil cover prediction with various amounts and types of crop residue, *Trans. Am. Soc. Agric. Eng.*, 25, 1333–1337.

Hassink, J., G. Leffink, and J. A. van Veen, 1991. Microbial biomass and activity of a reclaimed-polder soil under a conventional or a reduced-input farming system, *Soil Biol. Biochem.*, 23, 507–513.

Hoitink, H. A. J., T. Inbar, and M. J. Boehm, 1991. Status of compost amended potting mixes naturally suppressive to soilborne diseases of floriculture crops, *Plant Dis.*, 869–873.

Horwath, W. R., L. F. Elliott, and D. B. Churchill, 1995. Mechanisms regulating composting of high carbon to nitrogen ratio grass straw, *Compost Sci. and Utilization*, 3, 22–30.

Hunt, H. W., 1978. A simulation model for decomposition in grasslands, in *Grassland Simulation Model*, G. S. Ginnis, Ed., Springer-Verlag, New York, 155–183.

Iritani, W. M. and C. Y. Arnold, 1960. Nitrogen release of vegetable crop residues during incubation as related to their chemical composition, *Soil Sci.*, 89, 74–82.

Janzen, H. H. and R. M. N. Kucey, 1988. C, N, and S mineralization of crop residues as influenced by crop species and nutrient regime, *Plant Soil*, 106, 35–41.

Jawson, M. D. and L. F. Elliott, 1986. Carbon and nitrogen transformations during wheat straw and root decomposition, *Soil Biol. Biochem.*, 18, 15–22.

Knapp, E. B., L. F. Elliott, and G. S. Campbell, 1983a. Microbial respiration and growth during the decomposition of wheat straw, *Soil Biol. Biochem.*, 15, 319–323.

Knapp, E. B., L. F. Elliott, and G. S. Campbell, 1983b. Carbon, nitrogen and microbial biomass interrelationships during the decomposition of wheat straw: a mechanistic simulation model, *Soil Biol. Biochem.*, 15, 455–461.

Linn, D. M. and J. W. Doran, 1984. Effect of water-filled pore space on carbon dioxide and nitrous oxide production in tilled and nontilled soils, *Soil Sci. Soc. Am. J.*, 48, 1267–1272.

Lynch, J. M. and L. F. Elliott, 1997. Bioindicators: perspectives and potential value for landusers, researchers and policy makers, in *Biological Indicators of Soil Health and Sustainable Productivity*, C. Pankhurst, B. Doube, and V. Gupta, Eds., CAB International, Wallingford, U.K., 79–96.

Lynch, J. M., F. B. Ellis, S. H. T. Harper, and D. G. Christian, 1981. The effect of straw on the establishment and growth of winter cereals, *Agric. Environ.*, 5, 321–328.

Martin, J. P. and K. Haider, 1986. Influence of mineral colloids on turnover rates of soil organic carbon, in *Interactions of Soil Minerals with Natural Organics and Microbes*. P. M. H. M. Schnitzer, Ed., Soil Science Society of America Spec. Publ. 17, Madison, WI, 283–304.

McCalla, T. M. and T. T. Army, 1961. Stubble mulch farming, *Adv. Agron.*, 13, 125–196.

Nelson, E. B. and C. M. Craft, 1992. Suppression of dollar spot on creeping bentgrass and annual bluegrass turf with compost-amended topdressings, *Plant Dis.*, 76, 954–958.

Papendick, R. I., 1991. *Report on International Conference on the Assessment and Monitoring of Soil Quality*, Washington State University Press, Pullman, 38.

Papendick, R. I. and D. E. Miller, 1977. Conservation tillage in the Pacific Northwest, *Soil Water Conserv.*, 32, 49–56.

Papendick, R. I. and J. F. Parr, 1992. Soil quality—the key to sustainable agriculture, *Am. J. Alter. Agric.*, 7, 2–3.

Parr, J. F. and R. I. Papendick, 1978. Factors affecting the decomposition of crop residues by microorganisms, in *Crop Residue Management Systems*, W. R. Oshwald, Ed., American Society of Agronomy, Madison, WI, 101–129.

Reiners, W. A., 1968. Carbon dioxide evolution from the floor of three Minnesota forests, *Ecology*, 44, 471–483.

Reinertsen, S. A., L. F. Elliott, V. L. Cochran, and G. S. Campbell, 1984. Role of available carbon and nitrogen in determining the rate of wheat straw decomposition, *Soil Biol. Biochem.*, 16, 459–464.

Rovira, A. D. and E. L. Greacen, 1957. The effect of aggregate disruption on the activity of microorganisms in the soil, *Aust. J. Agric. Res.*, 8, 659–673.

Schippers, B., A. W. Bakker, and P. A. Bakker, 1987. Interaction of deleterious and beneficial rhizosphere microorganisms and the effect of cropping practices, *Annu. Rev. Phytopathol.*, 25, 339–358.

Smith, J. L. and L. F. Elliott, 1990. Tillage and residue management effects on soil organic matter dynamics in semiarid regions, *Adv. Soil Sci.*, 13, 69–87.

Sommers, L. E., C. M. Gilmour, R. E. Wildung, and S. M. Beck, 1981. The effect of water potential on decomposition processes in soils, in *Water Potential Relations in Soil Microbiology*, J. F. Parr, W. R. Gardner, and L. F. Elliott, Eds., Soil Science Society America Spec. Publ. 9, Madison, WI, 97–117.

Stott, D. E., 1991. RESMAN: A tool for soil conservation education, *J. Soil Water Conserv.*, 46, 332–333.

Stott, D. E. and J. B. Rogers, 1990. RESMAN: A Residue Management Decision Support Program, Public domain software, NSERL Publ. 5, 266 kb, USDA Agricultural Research Service National Soil Erosion Research Laboratory, West Lafayette, IN.

Stott, D. E., L. F. Elliott, R. I. Papendick, and G. S. Campbell, 1986. Low temperature or low water potential effects on the microbial decomposition of wheat residue, *Soil Biol. Biochem.*, 18, 577–582.

Stott, D. E., B. L. Stuart, and J. R. Barrett, 1988. Residue management decision support system, American Society of Agricultural Engineers, Microfiche Collect. Paper No. 88-7541, St. Joseph, MI.

Stott, D. E., H. F. Stroo, L. F. Elliott, R. I. Papendick, and P. W. Unger, 1990. Wheat residue loss from fields under no-till management, *Soil Sci. Soc. Am. J.*, 54, 92–98.

Stott, D. E., E. E. Alberts, and M. A. Welty, 1995. Plant residue decomposition and management, in USDA Water Erosion Prediction Project (WEPP): Hillslope Profile and Watershed Model Documentation, National Soil Erosion Research Laboratory Report No. 10, West Lafayette, IN (Software documentation).

Stroo, H. F., K. L. Bristow, L. F. Elliott, R. I. Papendick, and G. S. Campbell, 1989. Predicting rates of wheat residue decomposition, *Soil Sci. Soc. Am. J.*, 53, 91–99.

Summerall, B. A. and L. W. Burgess, 1989. Decomposition and chemical composition of cereal straw, *Soil. Biol. Biochem.*, 21, 551–559.

Suslow, T. V. and M. N. Schroth, 1982. Role of deleterious rhizobacteria as minor pathogens in reducing crop growth, *Phytopathology*, 72, 111–115.

Taylor, S. E. and O. J. Sexton, 1972. Some implications of leaf tearing in *Musaceae*, *Ecology*, 53, 143–149.

Thomsen, I. K., 1993. Nitrogen uptake in barley after spring incorporation of ^{15}N-labeled Italian ryegrass into sandy soils, *Plant Soil*, 150, 193–201.

Tisdall, J. M. and J. M. Oades, 1982. Organic matter and water-stable aggregates in soils, *J. Soil Sci.*, 33, 141–163.

U.S. Department of Agriculture, Soil Conservation Service and the Equipment Manufacturers' Institute, 1992. Estimates of Residue Cover Remaining after Single Operation of Selected Tillage Machines, USDA-NRCS, Washington, D.C.

van Veen, J. A. and E. A. Paul, 1981. Organic carbon dynamics in grassland soils. 1. Background information and computer simulation, *Can. J. Soil Sci.*, 61, 185–201.

Wiant, H. V., 1967. Influence of temperature on rate of soil respiration, *J. For.*, 65, 489–490.

Wieder, R. K. and G. E. Long, 1982. A critique of the analytical methods used in examining decomposition data obtained from litter bags, *Ecology*, 63, 1636–1642.

Wiegert, R. G. and F. C. Evans, 1964. Primary production and disappearance of dead vegetation on an old field in southeastern Michigan, *Ecology*, 45, 49–63.

Witkamp, M. and J. S. Olson, 1963. Breakdown of confined and nonconfined oak litter, *Oikos*, 14, 130–147.

Zaccheo, P., L. Crippa, and P. L. Genevini, 1993. Nitrogen transformation in soil treated with ^{15}N-labeled dried or composted ryegrass, *Plant Soil*, 148, 193–201.

5 Conservation Cropping Systems and Their Management

Clyde L. Douglas, Jr., Peggy M. Chevalier, Betty Klepper, Alex G. Ogg, Jr., and Paul E. Rasmussen

CONTENTS

5.1 INTRODUCTION

Techniques available for improving soil and water conservation fall into two broad categories: structure or landscape modification and management of soil surfaces and crop residues. This chapter addresses the second category, where crop rotation, tillage, residue management, and surface cloddiness are used for conservation in viable agricultural systems.

The chapter presents general principles used to develop management packages for production systems in the dryland areas of the Pacific Northwest (PNW). This

TABLE 5.1
Criteria That Define Agronomic Zones for the PNW East of the Cascade Mountains

				Criteria		
Zone	Name	Estimate Cultivated Acres[a]	Description	GDD, Fdd[b]	Soil Depth, in.	Annual Precipitation, in.
1	Mountain	—	Cold-moist	<1260	All	>16
2	Annual crop	1,778,000	Cool-moist	1260–1800	All	>16
3	Annual crop/fallow transition	684,000	Cool-deep-moderately dry	1260–1800	>40	14–16
4	Dry shallow	1,126,000	Cool-shallow-dry	<1800	<40	10–16
5	Dry deep	4,705,000	Cool-deep-dry	<1800	>40	10–14
6	Irrigated	1,317,000	Those parts of Zones 1-5 that are irrigated			

[a] From Smiley, 1992.

[b] Fdd = Fahrenheit degree days.

Adapted from Douglas et al., 1990.

region is especially suitable for testing soil and water conservation techniques and for defining boundaries of application because the area is relatively large, with gradual, well-defined trends in climate, soil properties, and elevation (USDA-SCS, 1981). The chapter will be organized by land-use zones where soil depth, air temperature, and precipitation divide the intermountain PNW into regions of similar agricultural management systems (Douglas et al., 1988; 1990; 1992). Table 5.1 summarizes criteria delineating these zones. By examining how conservation tools succeed or fail as one moves across agronomic zones, it is possible to draw general conclusions about limits of application of currently available practices.

5.2   AGRONOMIC ZONES FOR THE INTERMOUNTAIN PACIFIC NORTHWEST

A region including the portion of Oregon and Washington east of the Cascade mountains along with Idaho is divided into six dryland agronomic zones as shown in Figure 5.1 (Douglas et al., 1990). Zones range from a cold, moist, high-elevation area that once supported mountain forests (Zone 1) to an arid, hot, low-elevation zone (Zone 5) that was once a shrub-steppe (Douglas et al., 1992). Temperature, soil depth, and precipitation have been used to delineate these zones, except for Zone 6, which is based only on irrigation and will not be discussed. Because the climatic factors are correlated with elevation, lower-numbered zones generally apply to higher-elevation lands. Zone features, such as common crops, rotations, soil properties, and wheat yield expectations, are described in Table 5.2.

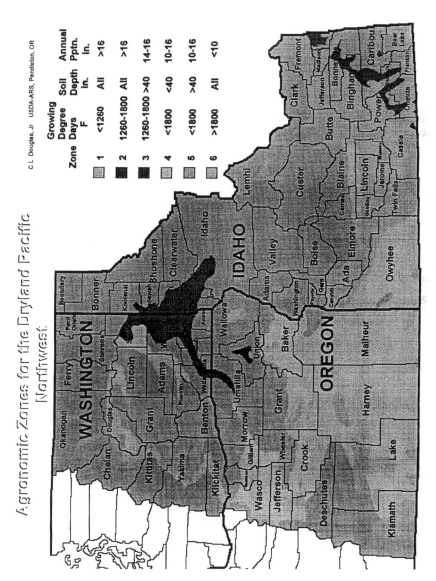

FIGURE 5.1   Map of the PNW east of the Cascade mountains showing the agronomic zone distribution.

TABLE 5.2
Agronomic Zones of the PNW East of the Cascade Mountains and Their Principal Agronomic Features

Zone	Parameters	Main Crops	Alternate and Minor Crops	Rotations	Average WW Yields, bu/ac	Soil Organic Matter, %	Water-Holding Capacity, in./ft.	Major Nutrient Deficiencies	Water Erosion Potential
1	Cold, moist	WC, SC, hay, pasture, grass seed	Vegetables, berries. lentils, canola	Annual Crop, BG(5–6 yr)/SB or DL/WW	70–90	4+	2.2–2.6	P, lime, B	High
2	Cool, moist	WC, SC, SDL. grass seed, canola	Canola	WW/DL, WW annually WW/SB/DL	80–120	3–4	2.0–2.4	N, S, P	High
3	Cool, deep soil, moderately dry	WW, SP (green, dry), lentils, SC. F	Canola, red lentils	WW/SP (green or dry), WW/F, WW/SC/F, annual WW	60–80	2–3	1.8–2.2	N, S	High
4	Cool, shallow soil, dry	WW, SB	SC, especially barley	WW/F, annual SB or after F, WW/SC/F	30–40	<1.5	1.6–2.0	N	High
5	Cool, deep soil, dry	WW	Canola, mustards, WC	WW/F, WW/SC/F, WW/F/F	40–60	<1.5	1.6–2.0	N	High

Abbreviations: BG = bluegrass, SP = spring peas, SB = spring barley, SC = spring cereal, SDL = spring dry legume, WW = winter wheat, WC = winter cereal. DL = dry legume, F = fallow, N = nitrogen, P = phosphorus, S = sulfur.

5.2.1 CLIMATE, SOIL, AND EROSION POTENTIAL

The latitude of the region ranges from 42 to 49° north latitude and daylength changes from about 16 h in summer to 8 h in winter (List, 1968). The region is characterized by a winter rainfall pattern with over 60% of the precipitation falling between November 1 and May 1, mostly as low-intensity rain or as snow (Molnau et al., 1987; Douglas et al., 1988). Precipitation increases with increasing elevation, from 10 in. near the Columbia River (350 ft elevation) to over 24 in. in the Blue Mountains of Washington and Oregon (3300 ft elevation). The winter precipitation pattern and deep soils lead to unusually good water storage for crop use compared with that in the central Great Plains of the U.S. (Ramig et al., 1983). The sole exception to a winter pattern occurs in southeastern Idaho (5000 to 6500 ft elevation), where an annual precipitation of 12 to 14 in. is more evenly distributed over the year and much of the water from summer rains is lost to evaporation.

Winter temperatures are mild compared with the midcontinent at these latitudes because the maritime effect of the Pacific Ocean extends far inland. At elevations greater than 2000 ft, winter temperatures frequently dip below 32°F and summer temperatures rarely rise above 90°F. At elevations below 2000 ft, winter temperatures are usually mild with daytime maximums often above 32°F and summer temperatures can reach 100°F. Temperatures are cooler at higher elevations and the northern part of the region is 3 to 6°F cooler than the southern part. Length of growing season decreases with increasing elevation.

Temperature variation across the region causes different rates of growing degree days (GDD) accumulation in the various zones (Table 5.1). Because most cool-season crops have a minimum growth temperature near the freezing point, GDD for crop growth was calculated with a base temperature of 32°F (Rickman et al., 1985). Cumulative GDDs for a site were calculated by summing average daily air temperatures for the period of crop development. For example, a day that has a maximum temperature of 78°F and a minimum temperature of 50°F would contribute $((78 + 50)/2) - 32$ or 32 GDD to the sum. Daily sums <32 were considered to be zero. For zone comparison, GDDs from January 1 through May 31 were summed (Table 5.1). This period was selected because many phenologically important events occur in the wheat crop during this time and areas too cold for production can be readily delineated. The January to June GDD accumulation in the mountain zone (Zone 1) is less than 1260. The value is between 1260 and 1800 for the other zones (Table 5.1). Cold temperatures and short growing seasons limit types of crops that can be grown in Zone 1. The remaining zones have sufficient GDDs for either winter or spring crops.

Intermittently frozen surface soils, common in much of the area, predispose soils on sloping land to excessive soil erosion (Hanson et al., 1988; McCool et al., 1987; McCool, 1990). In contrast to the eastern U.S., where energy from falling raindrops is the primary erosion mechanism, rain or snowmelt on frozen soil is the major cause of soil erosion in the PNW (Molnau et al., 1987; Kok and McCool, 1990 a, b). When a warm weather front passes over frozen soil, surface soils thaw but a lens of frozen soil, which is a barrier to water infiltration, remains in the profile (Zuzel and Pikul, 1987; Formanek, 1987). When surface thawing is combined with rain or snowmelt, additional water accumulates over the ice layer and supersaturates surface soil, causing it to lose strength and flow downhill. Steepness of the terrain increases the severity of erosion. Some farmlands in the Palouse have slopes greater than 50%.

The probability of a damaging erosion event can be decreased by surface residues. Even though frost does not penetrate as deeply when surface residue is present on the soil surface, thawing is more gradual and the potential for rapid thawing is less than when residue is not present (Pikul et al., 1986). Surface residues also effectively slow the flow of water over the surface, allowing more time for infiltration into the soil, effectively reducing erosion potential.

Most water erosion occurs in late winter (Istok et al., 1987). Annually cropped areas will have some protection at this time if sufficient residues from the previous crop remain on the surface. In fallow/wheat areas, the second winter of the rotation is critical because the winter wheat crop is too small to provide much protection and crop residues available for erosion protection are often minimal due to decomposition and tillage. Also, the soil profile may be nearly full from the previous winter's precipitation. Thus, management strategies must focus on maximum protection during late winter when erosion potential is maximum.

Generally, slope length and steepness are greatest for the Mountain and Annual Crop Zones, and water erosion potential decreases from Zone 1 to Zone 5. This decrease results from a combination of lower precipitation and less steep slopes at lower elevations. Exceptions to this trend occur in the 15- to 18-in. precipitation region of Washington and Oregon where erosion is more severe than in wetter or drier parts of the state (Roe and Rogers, 1987). Soils in these areas have relatively low organic matter and land is usually managed in a wheat/fallow, wheat/spring grain/fallow, or a wheat/pea rotation, which leaves little residue for soil protection in the wheat crop year. Much of this erosion-susceptible region is at an elevation with frequent freeze–thaw cycles.

The entire region has been exposed to wind erosion and redeposition of soil as a natural geologic process. Excessive tillage has accelerated soil movement in most parts of the region. Wind erosion is prevalent in all zones, but is especially severe in Zones 4 and 5.

Most soils in the PNW are aeolian in origin and are geologically and pedologically young. Cultivated soils in the region have primarily silt loam surface textures with a trend toward fine sandy loam at lower elevations and silty clay loam at higher elevations. Thus, water-holding capacity increases somewhat with higher precipitation, a trend accentuated by the higher levels of organic matter at higher elevation. Most soils in this area contain volcanic ash from recent activity in the Cascade mountains. Most have high native fertility for production of winter and spring cereals and legumes. Elements, such as potassium and phosphorus, that are primarily released from the inorganic fraction of soil are normally not deficient. Elements mineralized from the organic fraction of soil, such as nitrogen and sulfur, are generally deficient for optimum cereal yield (Rasmussen and Rohde, 1988; Rasmussen and Douglas, 1992). Increasing yields and continuing soil erosion have drastically increased the need for nitrogen and sulfur fertilization over the past 60 years (Rasmussen et al., 1989).

5.2.2 PRIMARY CROPS, ALTERNATE CROPS, AND DOMINANT CROP ROTATIONS

Historically, crops have been chosen primarily for yield, disease resistance, and readily available markets. Federal crop subsidy programs have also been an important factor in crop selection and land use. Table 5.2 lists the main crops and alternatives for each PNW zone. Short-season crops dominate because of the dearth of summer rain and the relatively short frost-free period. Winter cereals are grown in all zones, but are less important in Zone 1 because of the long winter snow cover. Spring cereals, crucifers, and legumes are used in Zones 1, 2, and 3, either as an annual crop or in rotation with a winter crop (Auld et al., 1987; Murray et al., 1987). Zone 1 is the only zone with significant arable dryland devoted to hay, pasture, or horticultural crops. In this zone, perennial crops cover steep slopes and wet soils.

Primary and alternate crops are rotated because crop rotation is the single most important disease-control method available (Cook and Veseth, 1991). This is a combination of cultural and biological control because most fungi that attack a specific crop cannot live in soil in the absence of the host for more than 2 years. Restricting wheat and barley in the same field to only once in 3 years will control or suppress soilborne diseases such as *Fusarium* root rot, take-all, strawbreaker foot rot, and *Cephalosporium* stripe (Mathre, 1982; Wiese, 1987).

Although some root pathogens such as *Rhizoctonia* and *Pythium* can live on roots of many different plants, yield losses can be reduced by rotating cereal crops with noncereal crops or fallow. *Pythium* can also be suppressed by planting clean seed not more than 1 year old and by planting early (Cook and Veseth, 1991). Metalaxyl as a seed treatment will also suppress *Pythium* and provide protection for developing seedlings (Wiese, 1987). *Rhizoctonia* is suppressed by soil disturbance, and is more of a problem in no-till than in conventionally tilled fields. *Rhizoctonia* can also be suppressed by eliminating the "green bridge" (bridging of green plants from one crop to the next) of volunteer crops and weeds that grow during periods between crops (Smiley et al., 1992). Late, shallow seeding and promotion of vigorous root growth with fertilizer reduce severity of *Rhizoctonia* (Mathre, 1982; Wiese, 1987). Planting disease-resistant cultivars is an effective method for control of snow mold, dwarf bunt, rusts, and strawbreaker foot rot (Wiese, 1987). Fungicides are available to aid in control of strawbreaker foot rot, rusts, and snowmold (Mathre, 1982; Wiese, 1987).

Increased efforts in erosion control and greater attention to sustainability of productive capacity have led to renewed interest in different rotations and new crops. Zones with higher precipitation are more amenable to annual cropping and thus to more complex crop rotations than are zones with lower precipitation (Table 5.2). Generally, crop rotation options decrease with lower precipitation (Bezdicek et al., 1987; Macnab and Ramig, 1987).

Soil depth is also a factor in the choice of cropping systems (see Table 5.1). Shallow soils (Zone 4) are prone to erosion in a cereal/fallow rotation since winter precipitation normally fills the soil profile every year. Deep soil profiles are often not filled by 1 year of precipitation. Thus, cereal/fallow rotations are recommended in Zone 5 but not in Zone 4. However, the most common rotation in Zone 4 during the last 30 to 50 years has been wheat/fallow.

5.2.3   YIELD POTENTIALS

Because of winter rainfall pattern, wheat can be produced with less annual precipitation in the PNW than in other parts of the world. The entire region has a near-perfect climate for growing winter wheat. Significantly more of the precipitation that falls in cooler months is stored for plant use compared with equal amounts of precipitation in the midwestern U.S., where summer rain predominates. Annual cereal yields generally increase as annual precipitation increases from Zone 5 to Zone 1, because water is the main factor limiting production (see Table 5.2). The exception is Zone 1, where production is limited by cold temperature rather than lack of water. Snow cover is desired in all zones to help protect crops against winter injury.

5.3   MANAGEMENT OF DRYLAND CROPPING SYSTEMS

5.3.1   THE MOUNTAIN ZONE—ZONE 1

The relatively high precipitation in this previously forested zone allows a greater diversity of options than in other zones. The main factors limiting crop selection are cool temperatures and a short growing season. Cold, wet springs limit both planting and emergence of spring crops because soils are difficult to work and plant growth is slow in cold and/or waterlogged conditions.

Soils are generally fertile. Organic matter levels of 4% provide a good supply of nitrogen through mineralization. Cold soil restricts root exploration of deeper parts of the soil profile and reduces the proliferation of roots at all depths. Phosphorus, which does not move readily in soils, must be intercepted by roots to be absorbed. Decreased root proliferation leads to lower plant uptake of available phosphorus, especially in early spring. Some soils in Zone 1 are of granitic origin and have a high phosphorus-fixing capacity, which restricts phosphorus available to plants.

Soil pH has been lowered during the past 20 to 30 years by natural leaching and increased use of acid-forming fertilizers (Mahler and Harder, 1984; Mahler et al., 1985; Rasmussen and Rohde, 1989). Liming is required for maximum crop production in the parts of the region that have historically received the most fertilizer (Mahler and McDole, 1985, 1987).

Because of high precipitation, both annual cropping and use of perennials such as pasture or hay are common in Zone 1. Most tilled land is in rotations such as winter wheat/spring cereal, annual spring cereal, or bluegrass/spring barley/winter wheat. Cool-season horticultural crops, such as onions, garlic, and strawberries, are produced as well as Christmas trees. Long-term (5 to >7 year) plantings of bluegrass (*Poa pratensis*) are grown for seed. Spring rapeseed (*Brassica napus* and *B. campestris*) has the potential to replace spring cereals in present rotations and may help break pest cycles in continuous cereal-cropping systems (Kephart et al., 1988).

Conservation tillage practices leave surface residues which act as a mulch to retard evaporative water loss and to slow heat exchange at the soil surface. Thus, soil conditions are generally cooler and wetter for spring operations under conservation tillage than with plowed soils.

Several plant diseases are associated with the cold, wet conditions in this zone. Cold wet soil at planting promotes attack of young seedlings by soil borne root pathogens such as *Pythium*. Long burial under snow favors snow mold (*Typhula* spp.). Dwarf bunt (*Tilletia controversa*) attacks wheat seedlings during winter. Stripe rust (*Puccinia striiformis*) and barley yellow dwarf are prevalent on wheat in this zone (Wiese, 1987; Cook and Veseth, 1991).

Weeds are difficult to control in this zone because of the continual emergence of new seedlings after each rain. Especially prevalent weeds are wild oat (*Avena fatua*) and Italian ryegrass (*Lolium multiflorum*), favored by wet soils and frequent use of spring cereals in the rotation. Wild oats can be controlled with preplant-incorporated applications of triallate or with postemergence applications of diclofop, imazamethabenz, and difenzoquat. Cultural practices such as high seeding rates can reduce wild oat competition in spring barley (Barton and Thill, 1991). Numerous broadleaf weeds occur, but most are controlled relatively easily with herbicides. Exceptions are catchweed bedstraw (*Galium* spp.), ivyleaf speedwell (*Veronica hederifolia*), and mayweed chamomile (*Anthemis cotula*), which are best controlled in the early seedling stage.

Cereal insect pests are few. Wireworms (*Limonius* spp.) can cause losses in cereal crops. Discontinuance of the use of persistent chlorohydrocarbons to control wireworms has increased populations over the last several years. Lindane is at present the most effective seed treatment for wireworm control.

Some soils are somewhat poorly drained because of texture changes between the topsoil and the subsoil that perches water during wet periods. Landscapes are often steep and water erosion potential is high. Wind erosion is minimal. Erosion control practices include crop residues and/or green cover on the surface, rough plowing, and grass rotations with no-till dry legume or cereal planting into grass sod. These practices have been adopted with success in many parts of the zone. Factors limiting widespread adoption include cold temperatures that reduce fall green cover growth in some high-mountain areas. Field management is difficult because of spots in fields that are not yet ready to work when the rest of the field is at the right moisture condition for tillage. Terraces, contour slopes, and divided slopes are also used in some areas of the zone.

Much of the land in this zone qualifies as "highly erodible" because of slope steepness and length and inherent soil erodibility. Some of the highly erodible land in Zone 1 was put into the Conservation Reserve Program (CRP) in the mid-1980s. Grass or grass–legume mixtures were planted at this time to establish vegetative cover without grazing or other disturbance. The U.S. government guaranteed a specific dollar amount per acre for a period of 10 years. Current CRP regulations, released in February 1997, precluded extension of any CRP contracts. Producers with contracts ending in 1997 have to reenroll under these new regulations. Enrollment can be for 10 to 15 years if eligibility requirements are met. Enrollment is limited to 36 million acres and only the most "environmentally sensitive" lands will be considered. Payment rates will be determined based on county average dryland cash or cash rent equivalent rental rates with adjustments for site-specific, soil-based productivity factors. There is also a provision for payment, not to exceed $5 per acre per year, to perform maintenance obligations.

In summary, cold, wet conditions in the mountain zone limit the time when soils can be tilled, favor soilborne root pathogens, reduce winter crop yield potential, promote certain weeds, and put early-planted spring crops at a disadvantage. However, relatively high fertility and available soil moisture increases yield potential and increases the diversity of crops that can be planted. Conservation tillage practices increase the coldness and wetness of soils in the spring.

5.3.2   THE ANNUAL CROP ZONE—ZONE 2

The steep hills of the Palouse and Nez Perce Prairies (USDA-SCS, 1981) compose most of Zone 2. This area is cool and relatively moist with deep soils. Soil erosion potential is very high and some hilltops are severely eroded with no remaining topsoil (Huggins and Pan, 1991). Soil nutrient deficiencies for high levels of wheat production include nitrogen, sulfur, and phosphorus. Phosphorus deficiency has been increasing as topsoil rich in organic matter is eroded (Hopkins and Pan, 1987). Legumes sometimes require molybdenum and boron.

Relatively high organic matter levels (3 to 4%) and good water-holding capacities of uneroded soils provide ideal conditions for overwinter soil water storage. However, steepness of terrain, inherent erodibility of soil, and frequent occurrence of rain or snowmelt on frozen soil require growers to implement erosion control practices (Mulla, 1986; Zuzel et al., 1986; Young et al., 1990). Most farms are planted somewhat on the contour. Strip-cropping and divided slopes have become more common in recent years. Terraces have been used to a limited extent. Wind erosion in this zone is minor compared with water erosion.

Some of the most productive wheat lands in the nation are located in this zone. The most common rotation is winter wheat/dry legume, but recropping of cereals is also prevalent. An increasing number of growers are using 3-year rotations, such as winter wheat/spring cereal/dry legume, to break up pest cycles associated with 2-year rotations and continuous cereals. Winter rapeseed (Kephart and Murray, 1990) is an alternate crop which may increase in acreage if a local processing plant is built and if winterkill problems are overcome. Rotations including bluegrass exist where spring legumes can be no-tilled into the bluegrass sod. This is usually followed by no-till winter wheat and then several years under a minimum tillage management, which either consists of two to three cycles of spring barley/spring legume/winter wheat or three to four cycles of spring lentils/winter wheat, before returning to 7 to 10 years of bluegrass.

Conservation practices most commonly used are rough moldboard plowing or chisel plowing in the fall to maintain some surface residue and increase surface roughness. Residues are abundant from cereal crops but not from legume crops. Excessive residue from cereal crops interferes with the proper functioning of seeding equipment. Fire is occasionally used to eliminate residue in no-till, but burning is discouraged because the long-term effect of repeated burning on soil quality is generally thought to be negative (Rasmussen et al., 1989). Insufficient legume residue can be a serious limitation to farmers who wish to use residues as a conservation practice on steep cropland, even with no-till seeding of winter cereals. A few farmers use very rough plowing after legume harvest to overcome the lack of residue after legume crops.

Downy brome (*Bromus tectorum*) is a major weed problem in continuous no-till. It has been listed most frequently by farmers as the reason that no-till systems fail (Papendick and Miller, 1977; Thill et al., 1987). Jointed goatgrass (*Aegilops cylindrica*) is increasing in all tillage systems, but more so under no-till. These two weeds are a serious problem in continuous wheat or in crop rotations where winter wheat is grown every other year. Prickly lettuce (*Lactuca serriola*), although easy to control with herbicides, increases in reduced and no-till systems. Catchweed bedstraw can be especially difficult to control. Wild oats, traditionally a weed problem in spring cereals, has become an increasing problem in winter cereals grown under reduced tillage. The increase in grassy weeds and certain broadleaf weeds in conservation tillage systems has been attributed to shallow seed burial and higher moisture in the surface soil in these systems (Appleby and Morrow, 1990).

Herbicides are widely used for weed control in small grains. Downy brome can be suppressed in conservation tillage systems with metribuzin applied postemergence or diclofop applied preemergence, but plant debris on the soil surface will reduce control. Downy brome can be decreased in reduced and no-till systems by using a 3-year rotation where winter wheat is grown only once. The interval between winter cereals should be extended to at least 3 years if jointed goatgrass is a problem, as there are no herbicides to control goatgrass selectively in wheat. Wild oats can be controlled in cereals grown under reduced tillage systems with a preplant incorporation of triallate and/or postemergence applications of diclofop, inazamethabenz, and difenzoquat. Most broadleaf weeds are controlled easily with mixtures of bromoxynil, dicamba, 2,4-D, MCPA, metribuzin, tribenuron, and trifensulfuron.

Insect pests of wheat include aphids such as bird cherry-oat aphid (*Rhopalosiphum padi*), English grain aphid (*Sitobion avenae*), rose grass aphid (*Metopolophium dirhodum*), greenbug (*Schizaphis graminum*), and Russian wheat aphid (*Diuraphis noxia*). Most aphids are more troublesome in spring wheat than in winter wheat, although greenbugs and Russian wheat aphids can be troublesome in early-planted winter wheat (Cook and Veseth, 1991). Most aphids can be controlled with insecticides such as disulfoton and dimethoate. Other insect pests in wheat include Hessian fly (*Mayetiola destructor*), wireworm (*Limonius species*), and wheat stem sawfly (*Cephus cinctus*). Hessian fly can be controlled by delaying planting of winter wheat until after mid-September or by planting Hessian fly-resistant spring wheats (Cook and Veseth, 1991).

Aphids can be troublesome in peas and lentils, and many fields are treated with insecticides each year. Aphids, such as the pea aphid (*Acyrthosiphon pisum*) also serve as vectors for pea enation mosaic virus (Hagedorn, 1984), which can completely destroy a pea crop. The pea weevil and pea leaf weevil can be destructive in peas, and fields are usually treated for these insects, especially for the pea weevil, which attacks the crop in the seedling stage. Management of pests in these crops will be covered in greater detail in Chapter 6.

Soilborne pathogens are common with considerable seedling damage from *Fusarium*, *Pythium*, and *Rhizoctonia*. Strawbreaker foot rot (*Pseudocercosporella herpotrichoides*) and *Cephalosporium* stripe (*Cephalosporium gramineum*) as well as several rusts (*Puccinia* spp.) are also important yield constraints (Line, 1987; Wiese et al., 1987). Principal control measures are rotations, which break the disease cycle (Pumphrey et al., 1987; Wiese et al., 1987; Cook, 1988, 1990; Cook et al., 1990; Cook and Veseth, 1991). *Aphanomyces* root rot (*Aphanomyces euteiches* Drechs) is one of the most destructive soilborne diseases of peas (Hagedorn, 1984). It can infect peas of any age and there is no cultural, chemical, or genetic control, other than eliminating peas and other host crops from rotations for 10 or more years. There are some commercial varieties that have measurable resistence to *A. euteiches* (Kraft et al., 1995). Inoculum has been lowered by plowing down green manure crops, such as oats (Muehlchen et al., 1990; Fritz et al., 1995). There is at present a cooperative effort between USDA-ARS and Oregon State University to evaluate integrated management systems for *A. euteiches* control in peas in the PNW. This effort is evaluating the combined effects of plant resistance, seed treatments, and cover crop plowdown.

5.3.3 THE ANNUAL CROP–FALLOW TRANSITION ZONE—ZONE 3

This zone is drier than Zone 2 and is found at an elevation from 1500 to 1800 ft on the western slopes of the Blue Mountains in Washington and Oregon and 5000 to 6500 ft on the western slopes of the Rocky Mountains in southeastern Idaho. In the higher precipitation areas of this zone, particularly in wetter years, there is sufficient precipitation for profitable annual cropping. Wind and water erosion potentials in this zone are high. Both wind and water erosion can occur during summer fallow. Water erosion can be severe in the winter of the crop year for winter cereals. Soils have a moderate water-holding capacity (1.8 to 2.2 in./ft) and lower organic matter content (2 to 3%) than soils in Zone 2. Winter wheat can be seeded after shallow-rooted or short-season crops such as green or dry peas. A green pea crop uses soil moisture only to about 36 in.; thus, some soil water from the pea-cropping year can be saved for the wheat-cropping year (Pumphrey et al., 1979). In the lower precipitation areas of this zone, and in drier years, growers plant winter wheat after fallow. Predominant rotations in this zone include winter wheat/spring cereal/fallow in Washington and winter wheat/spring legume in Oregon. However, this practice may increase soil erosion when soils are not able to hold the moisture received during the winter when the crop is growing. A few growers are experimenting with spring canola and spring red lentils. Annual recropped winter wheat is grown on a very limited basis. Federal farm programs have significantly influenced choices available in the past.

Major weed problems in Zone 3 are downy brome, jointed goatgrass, prickly lettuce, coast fiddleneck (*Amsinckia intermedia*), several mustards, prostrate knotweed (*Polygonium aviculare*), Russian thistle (*Salsola iberica*), and wild oat. Downy brome and jointed goatgrass are problems mainly in winter wheat, whereas wild oat is the main problem in spring cereals. Broadleaf weeds occur in both winter and spring cereals. Downy brome is more difficult to control in conservation tillage systems than in conventionally tilled systems under 2-year rotations or continuous wheat, because surface residues interfere with incorporation of herbicides and intercept herbicides applied as liquids. Most other weeds, except jointed goatgrass, can be controlled by proper selection and application of herbicides. Because of repeated and widespread use of sulfonylurea herbicides, populations of Russian thistle and kochia resistant to these herbicides have been inadvertently selected. Dicamba and 2,4-D must now be used with sulfonlyurea herbicides to ensure control of these weeds. Lack of adequate and safe weed control practices is a major constraint to the introduction of spring red lentils into this zone. Weeds such as coast fiddleneck, prostrate knotweed, and Russian thistle are the most troublesome.

Major diseases and insects affecting crops in this zone are similar to those in Zone 2. Russian wheat aphid in wheat and barley and pea aphid, pea weevil, and pea leaf weevil in peas are the major insect pests in this zone. Early-planted winter wheat is particularly susceptible to Russian wheat aphid and may require fall application of insecticide. Pea aphids and weevils are at present controlled with insecticides.

Erosion control practices include grass waterways, terraces, strip-cropping, and contour farming. Trashy fallow is often used for erosion control in drier areas. Utilization of minimum tillage to plant wheat is often done to control erosion. Tillage may consist only of a heavy-duty fertilizer applicator without prior tillage after legumes, or chisel plowing or rough moldboard plowing after spring or winter cereals (Veseth et al., 1992). In southeastern Idaho, chiseling is common to facilitate water intake into frozen soils (Massee and Siddoway, 1969).

Several farmers have adopted no-till production systems for this zone, although one or more crops in the rotation may be planted utilizing minimum tillage. One of the most effective rotations has been winter wheat/spring barley/chemical fallow. The use of glyphosate before planting spring barley and during the fallow year is a key component in this system. Initially, this system is heavily dependent upon herbicides to control weeds; but, after a few years, weed problems tend to decrease. Most other farmers still use mechanical tillage during the fallow year, although glyphosate is increasingly being used to delay primary tillage. Rod-weeding creates a potential wind erosion problem unless adequate surface residues and/or soil structure are maintained.

5.3.4 THE ARID ZONES—ZONE 4 (SHALLOW SOIL) AND ZONE 5 (DEEP SOIL)

Drier regions are divided into Zone 4, which has shallow soils that fill to capacity with water every year, and Zone 5, which is similar to Zone 4, except for having soils deep enough to hold the moisture collected during two winters. Soils in both zones have low organic matter (1 to 2%) but slightly less water-holding capacity (1.6 to 2.0 in./ft) than soils in Zone 3. Nitrogen is the primary nutrient limiting wheat production. Some soils in Zone 4 have caliche layers that restrict rooting. Lack of adequate water is the main yield-limiting factor in both zones. Water erosion potential is moderate to high. Wind erosion potential is high, especially when dust mulches are not protected by residue. Erosion control measures include standing stubble over winter, rough tillage, trashy fallow, and adequate surface residue at seeding. Low production limits residues available for erosion control; therefore, some farmers use soil surface roughness to help reduce soil erosion. Terraces, contour strips, and divided slopes are often used for erosion control.

Zone 4 could and should be cropped annually because the fallow period does not store additional water for the following crop and can result in significant water erosion. Most growers persist in using a wheat/fallow rotation because of its low risk. Those who crop annually, recrop to winter wheat or spring cereals or alternate between winter wheat and spring wheat. Spring red lentils are being evaluated as an alternate crop to spring cereals. Winter wheat yields in Zone 4 are limited to approximately 30 bu/acre because of limited soil water-holding capacity.

Winter wheat in Zone 5 is produced in a fallow/wheat rotation on soil water stored over 2 years. Only about 60 to 75% of the first winter's moisture is stored at the end of winter with most moisture deep in the profile (Ramig and Ekin, 1991). Occasionally, alternate crops such as winter or spring rapeseed, mustard, and spring cereals are grown. Most of the land is in a 2-year rotation, but some growers use winter wheat/spring crop/fallow rotations, or continuous spring cereals. A small number of growers use winter wheat/fallow/fallow in extremely dry areas. Extensive areas in both Zones 4 and 5 have been planted to grass in the CRP.

Principal diseases are *Fusarium* foot rot, *Cephalosporium* stripe, strawbreaker foot rot, rusts, barley yellow dwarf, *Rhiztonia* root rot, and, in the colder parts in northern Washington and Idaho and the southeastern part of Idaho, snow mold. Insecticides sometimes must be applied to control grasshoppers (*Melanoplus* spp.) and Russian wheat aphids.

Weeds are a serious problem in Zones 4 and 5 because of severe limitations on plant-available water. Common weeds are downy brome (*Bromus tectorum* L.), field bindweed, cereal rye, Russian thistle, several mustards, jointed goatgrass (*Aegilops cylindrica*), and kochia in some areas. All except jointed goatgrass, downy brome, and sometimes Russian thistle can be controlled by proper selection and application of herbicides. During the fallow year, weeds can be controlled with a field cultivator and rod-weeder. In addition, 2 or more years without winter cereals decreases problems with downy brome and jointed goatgrass. Granular triallate has been used to preplant in trashy fallow to control downy brome in winter wheat; however, it is relatively expensive. Granules are applied prior to the final rod-weeding and work their way through the crop debris to the soil surface to control grass as it germinates. Sulfonylurea-resistant Russian thistle and kochia are even more common than in Zone 3 and bromorynil and dicamba or bromorynil plus 2,4-D as a tank mix must be used to control these weeds.

5.4   CONCLUSIONS—PRINCIPLES AND RESEARCH NEEDS

Each of the five agronomic zones has different problems associated with its environment. Discussion will be organized by topic across zones to show how problems change across environments.

5.4.1   CROP RESIDUES

Cereal residues available for erosion control decrease markedly from wetter to drier zones. For every bushel of winter wheat grain produced, there will be about 80 to 100 lb of residue in the field at the end of harvest. For spring wheat, residues are less because grain yields are lower and residue-to-grain ratio is usually smaller (Table 5.3). Residue amounts for legumes are even lower than for spring cereals. Thus, winter wheat residue in Zone 2 can be more than twice that in Zone 4. Farmers in drier

TABLE 5.3

Average Grain Yields and Residue-to-Grain Ratios for Major Crops in Five Agroclimatic Zones (amounts of residue per acre are calculated from the product of the two numbers given)

Zone	Winter Wheat Yield, lb/ac	Winter Wheat Res/Grain[b]	Spring Wheat Yield, lb/ac	Spring Wheat Res/Grain	Winter Barley Yield, lb/ac	Winter Barley Res/Grain	Spring Barley Yield, lb/ac	Spring Barley Res/Grain	Peas[a] Green Yield, lb/ac	Peas[a] Green Res/Grain	Peas[a] Dry Yield, lb/ac	Peas[a] Dry Res/Grain	Lentils[a] Yield, lb/ac	Lentils[a] Res/Grain
1	4800	1.40[c]	3600	1.16	5400	1.13	3700	1.26	3500	1.3	1800	1.4	—	—
2	5400	1.33	4500	1.24	6000	0.95	5300	0.93	2000	1.4	2200	1.3	1600	1.7
3	4200	1.45	3300	1.33	5200	1.18	4700	1.00	—	—	1600	1.6	1000	2.0
4	2100	1.69	1500	1.69	1800	2.25	1700	1.76	—	—	—	—	—	—
5	3000	1.48	2400	1.40	3500	1.50	1900	1.60	—	—	—	—	—	—

[a] Information on dry pea and lentil yields from personal communication with F. Muehlbauer, USDA, ARS, Pullman, WA and green pea yields from personal communication with T. Darnell, Oregon State University Cooperative Extension, Milton Freewater, OR.

[b] Res/grain = residue to grain ratios (adapted from McClellan et al., 1987).

[c] Yield calculations based on 60 lb/bu for wheat and 48 lb/bu for barley.

zones often lack sufficient residue to control erosion, whereas those in wetter zones struggle with too much.

Use of residues for erosion control has the least impact on spring crop growth, where soils are well-drained, early spring temperatures are relatively warm, and water is limited. The residue mulch causes surface soils to remain cooler and wetter in spring. Heavy residues can be damaging to spring crops in wetter, cooler zones when early waterlogging and low temperatures are unfavorable for crop growth. Soil wetness can also limit tillage operations in Zones 1 and 2. For the wetter zones, effective erosion control practices that permit early entry into fields would be useful. Work is also needed to determine if "ridge tillage" that is successful in the midwestern U.S. could be adapted in any areas of the PNW drylands.

Residues will not be distributed uniformly across the field unless special harvesting equipment is utilized (Allmaras et al., 1985; Douglas et al., 1989). Residue distribution patterns behind combines become more critical in the wetter zones, where "chaff rows," and associated volunteers and weeds can increase the potential for root diseases (Cook et al., 1990). Management of large amounts of straw without heavy tillage or burning is indeed a challenge in Zones 1, 2, and 3. Residues for erosion control depend not only on the amount left at harvest but also on tillage and length of time that residues are subjected to environmental conditions that promote decay. Temperature and inherent nitrogen content of residues are critical to the decay process (Douglas and Rickman, 1992). Surface roughness can compensate for some residue requirements for erosion control.

Variable production potential and resource protection considerations within fields in PNW cropland need to be integrated into the development of "targeted management" for specific parts of fields. Fertilizer requirements, residues produced, pest management, residue management, and tillage operations for optimum soil water storage and erosion control are often different for different parts of the landscape. If these differences are recognized and appropriate practices applied across the field, there should be more efficient use of production inputs and greater resource protection than with the present system of applying practices uniformly across the landscape. Variable tillage and residue management practices may only apply to primary tillage operations, or differences may continue across the whole field until planting (Veseth et al., 1992). Strip-cropping and divided slope farming have potential for allowing more targeted management on variable cropland. In the future, precision farming using global positioning systems (GPS) will allow chemical applications tailored to the requirements of specific field sections. More research is needed on the best way to divide fields for efficient equipment operation as well as for residue and crop management. It is necessary to understand nutrient cycling patterns of different parts of the landscape (Fiez et al., 1994, 1995) and to devise methods for varying chemical, seed, and tillage operations. More-precise application of agricultural chemicals and nutrients will provide improved protection of surface and groundwater.

Residues increase water conservation, decrease wind and water erosion, and retard soil frost penetration in winter and heat penetration in summer (Douglas et al., 1987). Quantitative data to relate amounts of surface residue to each of these are needed. Process models must be developed to predict in any zone the effectiveness of residues over time as related to cropping and environmental variables. Little information is available on partial burial affects on residue effectiveness in erosion control. More work is needed to clarify the interaction between tillage and residue burial for different crops. Although residue management data are available for cereals, they are not commonly available for alternate crops such as legumes or mustards.

Equipment operations of all types are affected by residues. Design criteria for construction and operation of farm implements have been developed, especially with respect to seeding and fertilizer placement (Wilkins et al., 1987; Wilkins, 1988). Good standing establishment of wheat or other crops in high-residue situations needs to be improved, although several effective drills have been developed (Hyde et al., 1987). For very high residue situations, some type of drill attachment that clears residue from the crop row might be helpful.

In conservation tillage systems, seedlings are subject to soilborne pathogens as well as the lack of nutrient availability because of microbial activity associated with straw decomposition (Cook and Murray, 1987). Planting vigorous seed can obviate some of these difficulties (Hering et al., 1987). However, crop rotations effective in breaking up disease cycles and economically viable alternate crops have not yet been developed for the drier zones. Improvement in planting equipment and technology for seeding into cloddy seedbeds are needed where surface roughness can be used for additional erosion control.

5.4.2 CROPS

A major factor limiting the development of viable rotation systems, especially in Zones 4 and 5, is the need for more crop options. Limitations on rotational crops include lack of ready markets capable of absorbing the quantities produced; lack of herbicide labels for minor crops; lack of drought-hardy, disease- and insect-resistant competitive crops; and the need for livestock to be more effectively integrated into the system so more land can be profitably devoted to pasture or hay. A winter-hardy legume such as Austrian winter pea (*Pisum sativum* var. arvense) or other alternate crops would give more rotational options in Zones 2 and 3. Before 1997, government farm programs had a number of restrictions that limited planting flexibility needed for use of continuous cropping and crop rotations, without loss of commodity base. The Agricultural Market Transition Act (AMTA) of 1996 removed a number of these restrictions. Producers can now plant any crop except fruits (including nuts) and vegetables (excluding lentils, mung beans, and dry peas) on contract acreage on their farm. There is also no penalty for planting in excess of contract acreage.

An alternative crop to replace fallow, even if it were only a green manure crop, would help to keep soils in the drier parts of the region covered and protect them against wind and water erosion. Green manures have not been grown widely in the past because they use up precious water which farmers normally reserve for the following wheat crop. For every inch of water used by a cover crop, there is a 5 to 7 bu/acre loss of wheat yield (Leggett et al., 1974). Also, retaining water in the seed zone is very critical for crop establishment. As an alternate to green manure, periodic planting to grass 3 to 5 years might provide benefits in terms of erosion control and also increase soil fertility and manageability. Someday, modification of farm programs may encourage farmers in the drier zones to plant cover crops for erosion control.

Spring crops or cultivars that will germinate, emerge, and thrive in cold, wet soils would be useful in Zones 1 and 2. For those crops currently grown, new disease and insect resistance in cultivars would help increase rotation flexibility. For zones with significant winter damage potential, alternate crops are needed that have sufficient cold hardiness to survive winter and sufficient fall growth to provide ground cover overwinter. Spring wheat cultivars that yield well would be especially useful as an annual crop on shallow soils and in wetter areas to form an economically viable package in a 3-year or longer rotation.

Most farmers are aware of the benefits of rotation to control pests, but more research is needed to establish economically viable alternate crops, especially for Zones 4 and 5. Replacing fallow in drier zones would provide additional plant residues to help control soil erosion, and provide annual inputs to soil organic matter to help maintain soil tilth and fertility. Alternate green manure legume crops would also produce nitrogen for use by the following wheat crop and reduce the need for nitrogen fertilizer.

5.4.3 WEEDS, DISEASES, AND OTHER PESTS

Development of pest-resistant cultivars would decrease both production cost and potential problems associated with use of pesticides. Further work is certainly needed on effective use of chemicals, even if resistant cultivars are available, because of the probability of failure of crop genetic resistance as pests evolve. There is a need for improved cultivars, seed production techniques, and planting technology to increase competitiveness of crop seedlings against pests. A competitive seedling crop can effectively close the canopy and shade weeds or it can "outgrow" certain pathogens through rapid, vigorous growth.

Future pest control efforts include the need for better weed control in cereal crops. Special needs include the control of downy brome and jointed goatgrass in winter wheat in all zones and control of Russian thistle and kochia in winter and spring cereals in fallow in drier zones. Control practices in the drier zones will need to address the evolution and spread of herbicide-resistant weeds. There is a special need for better control of weeds in broadleaf crops such as spring red lentils and canola. Cereal cultivars resistant or tolerant to Russian wheat aphid, greenbug, *Cephalosporium* stripe, *Fusarium* root rot, and *Rhizoctonia* root rot would be welcome additions to the arsenal of tools farmers have available for pest control.

5.4.4 SOIL PROPERTIES

All crops and residue management technologies have implications for long-term effects on soil properties. For example, fertilization with ammonium-based nitrogen fertilizer has made surface soil more acid. This acidity reduces crop production, increases diseases such as *Cephalosporium* stripe, and increases certain weeds such as wild oat. Zones 1 and 2 are especially affected because they have received the most nitrogen fertilizer. Agronomic and economic evaluations of liming are needed in these areas.

The long-term effects on soil organic matter from practices such as burning straw or using different rotations in each zone need to be assessed. There is little quantitative data on impacts of burning stubble and then no-till planting into the burned stubble. Both the long-term effects on soil properties and the effectiveness of remaining stubble in controlling erosion need to be assessed before this practice has a place in the list of conservation options.

Previous work has shown that fallowing is the practice most destructive of soil organic matter content. Declines in organic matter levels are slower under any annual cropping scheme than under wheat/fallow (Rasmussen et al., 1989). If tillage-based fallow is to be eliminated, methods must be found to keep the soil covered and weed-free and to retard evaporation from the soil surface.

Wind erosion is a major concern in Zones 4 and 5. Alternatives to the dust mulch fallow, maintained with intensive tillage and residue incorporation, are needed to control erosion by either wind or water. The dust mulch does an excellent job of keeping moisture near the soil surface to permit early fall planting of cereals. Rod-weeding to maintain the dust mulch also effectively uproots young weeds. Replacing both benefits will be difficult.

5.4.5 TECHNOLOGY TRANSFER

Most farmers try to keep up with the latest changes in equipment. They listen carefully to information on new cultivars. They experiment with new practices that show promise for helping solve problems. However, up until now options were somewhat restricted by the Farm Program, and farmers had to be very careful to do nothing that would decrease their commodity base and payments. If erosion control were made more a part of the economic picture than it currently is, then some options and rotations would become more attractive. The single most important change that would benefit agriculture in the dryland PNW would be more rotation options. Research by integrated teams is urgently needed. Current information can be made available to growers through on-farm research and educational efforts as a part of the Solutions to Economic and Environmental Problems (STEEP) Program and other extension programs in the public and private sectors. Technology transfer to growers is very important to the continuing success of agricultural research in the region.

REFERENCES

Allmaras, R. R., C. L. Douglas, Jr., P. E. Rasmussen, and L. L. Baarstad, 1985. Distribution of small grain residues produced by combines, *Agron. J.*, 77, 730–734.

Appleby, A. P. and L. A. Morrow, 1990. The Pacific Northwest, in W. W. Donald, Ed., *Systems of Weed Control in Wheat in North America*, Weed Science Society of America, Champaign, IL, 200–232.

Auld, D. L., G. A. Murray, and F. V. Pumphrey, 1987. Alternate crops in conservation tillage systems, in L. F. Elliott, Ed., *STEEP—Conservation Concepts and Accomplishments*, Washington State University Press, Pullman, 137–157.

Barton, D. L. and D. C. Thill, 1991. Effects of cultural and chemical wild oat control on maximum economic yield of spring barley, *Proc. West. Soc. Weed Sci.*, 44, 79.

Bezdicek, D. F., C. S. Root, E. Kirby, D. Granatstein, L. F. Elliott, and R. I. Papendick, 1987. Tillage and cropping system alternatives for high rainfall areas, in L. F. Elliott, Ed., *STEEP—Conservation Concepts and Accomplishments*, Washington State University Press, Pullman, 409–418.

Cook, R. J., 1988. Interactions of tillage and soil management practices on the biological control of diseases and pests, in R. J. Summerfield, Ed., *World Crops: Cool Season Food Legumes, Proceedings of the International Food Legume Research Conference of Pea, Lentil, Faba Bean and Chickpea*, Spokane, WA, 6–11 July 1986, 649–660.

Cook, R. J., 1990. Diseases caused by root-infecting pathogens in dryland agriculture, *Adv. Soil Sci.*, 13, 214–239.

Cook, R. J. and T. D. Murray, 1987. Management of soilborne pathogens of wheat in soil conservation systems, in L.F. Elliott, Ed., *STEEP—Conservation Concepts and Accomplishments*, Washington State University Press, Pullman, 314–323.

Cook, R. J. and R. J. Veseth, 1991. *Wheat Health Management*, American Phytopathological Press, St. Paul, MN, 152 pp.

Cook, R. J., C. Chamswarng, and W.-H. Tang, 1990. Influence of wheat chaff and tillage on *Pythium* populations and *Pythium* damage to wheat, *Soil Biol. Biochem.*, 22, 939–947.

Douglas, C. L., Jr., and R. W. Rickman, 1992. Estimating crop residue decomposition from air temperature, initial nitrogen content, and residue placement, *Soil Sci. Soc. Am. J.*, 56, 272–278.

Douglas, C. L., Jr., R. E. Ramig, P. E. Rasmussen, and D. E. Wilkins, 1987. Residue management: small grains in the Pacific Northwest, *Crops Soils*, 39, 22–24.

Douglas, C. L., Jr., R. W. Rickman, J. F. Zuzel, and B. L. Klepper, 1988. Criteria for delineation of agronomic zones for the Pacific Northwest, *J. Soil Water Conserv.*, 43, 415–421.

Douglas, C. L., Jr., P. E. Rasmussen, and R. R. Allmaras, 1989. Cutting height, yield level, and equipment modification effects on residue distribution by combines, *Trans. ASAE*, 32, 1258–1262.

Douglas, C. L., Jr., D. J. Wysocki, J. F. Zuzel, R. W. Rickman, and B. L. Klepper, 1990. Agronomic Zones for the Pacific Northwest, PNW 354, A PNW Extension Publication, Oregon, Idaho, Washington, 8 pp.

Douglas, C. L., Jr., R. W. Rickman, B. L. Klepper, J. F. Zuzel, and D. J. Wysocki, 1992. Agroclimatic zones for dryland winter wheat producing areas of Idaho, Washington, and Oregon, *Northwest Sci.*, 66, 26–34.

Fiez, T. E., B. C. Miller, and W. L. Pan, 1994. Assessment of spatially variable nitrogen fertilizer management in winter wheat, *J. Prod. Agric.*, 7, 86–93.

Fiez, T. E., W. L. Pan, and B. C. Miller, 1995. Nitrogen use efficiency of winter wheat among landscape positions, *Soil Sci. Soc. Am. J.*, 59, 1666–1671.

Formanek, G. E., 1987. Effect of tillage and cropping system alternatives on runoff and soil erosion of the Pacific Northwest, in L.F. Elliott, Ed., *STEEP—Conservation Concepts and Accomplishments*, Washington State University Press, Pullman, 419–427.

Fritz, V. A., R. R. Allmaras, F. L. Pfleger, and D. W. Davis, 1995. Oat residue and soil compaction on common root rot (*Aphanomyces euteiches*) of peas in a fine-textured soil, *Plant Soil*, 171, 235–244.

Hagedorn, D. J., Ed., 1984. *Compendium of Pea Diseases*, American Phytopathological Society, St. Paul, MN, 57 pp.

Hanson, C. L., D. P. Burton, and M. Molnau, 1988. Soil Frost Distribution and Occurrence on a Mountainous Rangeland Watershed in Southwestern Idaho, Idaho Agricultural Experiment Station Research Bulletin No. 142.

Hering, T. F., R. J. Cook, and W.-H. Tang, 1987. Infection of wheat embryos by *Pythium* species during seed germination and the influence of seed age and soil matric potential, *Phytopathology*, 77, 1104–1108.

Hopkins, A. G. and W. L. Pan, 1987. Variation in P availability in a Palouse toposequence, in *Western Phosphate Conference Abstracts*, Corvallis, OR, March 15–16, 1987, 31–38.

Huggins, D. R. and W. L. Pan, 1991. Wheat stubble management affects survival, growth and yield of winter legumes, *Soil Sci. Soc. Am. J.*, 55, 823–829.

Hyde, G., D. Wilkins, K. Saxton, J. Hammel, G. Swanson, R. Hermanson, E. Dowding, J. Simpson, and C. Peterson, 1987. Reduced tillage seedling equipment development. in L. F. Elliott, Ed., *STEEP—Conservation Concepts and Accomplishments*, Washington State University, Pullman, 41–56.

Istok, J. D., J. F. Zuzel, L. Boersma, D. K. McCool, and M. Molnau, 1987. Advances in our ability to predict rates of runoff and erosion using historical climatic data, in L. F. Elliott, Ed., *STEEP—Conservation Concepts and Accomplishments*, Washington State University Press, Pullman, 205–223.

Kephart, K. D. and G. A. Murray, 1990. Dryland Winter Rapeseed Production Guide, University of Idaho, Bull. No. 715.

Kephart, K. D., M. E. Rice, J. P. McCaffrey, and G. A. Murray, 1988. Spring Rapeseed Culture in Idaho, University of Idaho Extension Bull. 681.

Kok, H. and D. K. McCool, 1990a. Quantifying freeze-thaw induced variability of soil strength, *Trans. ASAE*, 33, 502–506.

Kok, H. and D. K. McCool, 1990b. Freeze thaw effects on soil strength, in K. R. Cooley, Ed., *Proceedings, International Symposium—Frozen Soil Impacts on Agricultural, Range, and Forest Lands*, March 21–22, 1990. Spokane, WA, 70–76.

Kraft, J. M., V. A. Coffman, and T. J. Darnell, 1995. Pea common root rot control with cultivar resistance and biocontrol seed dressing, *Biol. Cult. Control Plant Dis.*, 10, 139.

Leggett, G. E., R. E. Ramig, L. C. Johnson, and T. W. Massee, 1974. Summer fallow in the Northwest, in Summer Fallow in the Western United States, Conservation Research Rep. 17, USDA, ARS, 110–135.

Line, R. F., 1987. Foliar diseases and conservation farming systems in the Pacific Northwest, in L.F. Elliott, Ed., *STEEP—Conservation Concepts and Accomplishments*, Washington State University Press, Pullman, 543–549.

List, R. J., 1968. *Smithsonian Meteorological Tables*, 6th rev. ed., Smithsonian Institution Press, Washington, D.C.

Macnab, A. W. and R. E. Ramig, 1987. Tillage and cropping system alternatives for low rainfall areas in the Pacific Northwest, in L.F. Elliott, Ed., *STEEP—Conservation Concepts and Accomplishments*, Washington State University Press, Pullman, 395–407.

Mahler, R. L. and R. W. Harder, 1984. The influence of tillage methods, cropping sequence, and N rates on the acidification of a northern Idaho soil, *Soil Sci.*, 137, 52–60.

Mahler, R. L. and R. E. McDole, 1985. The influence of lime and phosphorus on crop production in northern Idaho, *Commun. Soil Sci. Plant Anal.*, 16, 485–499.

Mahler, R. L. and R. E. McDole, 1987. Effect of soil pH on crop yield in northern Idaho, *Agron. J.*, 79, 751–755.

Mahler, R. L., A. R. Halvorson, and F. E. Koehler, 1985. Long-term acidification of farmland in northern Idaho and eastern Washington, *Commun. Soil Sci. Plant Anal.*, 16, 83–95.

Massee, T. W. and F. H. Siddoway, 1969. Fall chiseling for annual cropping of spring wheat in the Intermountain dryland region, *Agron. J.*, 61, 177–182.

Mathre, D. E., Ed., 1982. *Compendium of Barley Diseases*, American Phytopathological Society, St. Paul, MN, 78 pp.

McClellan, R. C., T. L. Nelson, and M A. Sporic, 1987. Measurements of residue-to-grain ratio and relative amounts of straw, chaff, awns and grain yield of wheat and barley varieties common to eastern Washington, in L. F. Elliott, Ed., *STEEP—Conservation Concepts and Accomplishments*, Washington State University Press, Pullman, 617–624.

McCool, D. K., 1990. Crop management effects on runoff and soil loss from thawing soil, in K. R. Cooley, Ed., *Proceedings, International Symposium—Frozen Soil Impacts on Agricultural, Range, and Forest Lands*, March 21–22, 1990. Spokane, WA, 171–176.

McCool, D. K., J. F. Zuzel, J. D. Istok, G. E. Formanek, M. Molnau, K. E. Saxton, and L. F. Elliott, 1987. Erosion processes and prediction for the Pacific Northwest, in L. F. Elliott, Ed., *STEEP—Conservation Concepts and Accomplishments*, Washington State University Press, Pullman, 187–204.

Molnau, M., J. F. Zuzel, J. D. Istock, K. E. Saxton, and D. K. McCool, 1987. Hydrology processes and prediction in the Pacific Northwest, in L.F. Elliott, Ed., *STEEP—Conservation Concepts and Accomplishments*, Washington State University Press, Pullman, 159–186.

Muehlchen, A. M., R. E. Rand, and J. L. Parke, 1990. Evaluation of crucifer green manures for controlling Aphanomyces root rot of peas, *Plant Dis.*, 74, 651–654.

Mulla, D. J., 1986. Distribution of slope steepness in the Palouse region of Washington, *Soil Sci. Soc. Am. J.*, 50, 1401–1405.

Murray, G. A., D. L. Auld, and F. V. Pumphrey, 1987. Alternate crops for Pacific Northwest rotation and tillage systems, in L.F. Elliott, Ed., *STEEP—Conservation Concepts and Accomplishments*, Washington State University Press, Pullman, 595–597.

Papendick, R. I. and D. E. Miller, 1977. Conservation tillage in the Pacific Northwest, *J. Soil Water Conserv.*, 32, 49–56.

Pikul, J. L., Jr., J. F. Zuzel, and R. N. Greenwalt, 1986. Formation of soil frost as influenced by tillage and residue management, *J. Soil Water Conserv.*, 41, 196–199.

Pumphrey, F. V., R. E. Ramig, and R. R. Allmaras, 1979. Field response of peas (*Pisum sativum* L.) to precipitation and excess heat, *J. Am. Soc. Hort. Sci.*, 104, 548–550.

Pumphrey, F. V., D. E. Wilkins, D. D. Hane, and R. W. Smiley, 1987. Influence of tillage and nitrogen fertilizer on *Rhizoctonia* root rot (bare patch) of winter wheat, *Plant Dis.*, 71, 125–127.

Ramig, R. E. and L. G. Ekin, 1991. When do we store water with fallow? in Columbia Basin Agricultural Research Spec. Rpt. 879. Oregon State Univ., Corvallis, 56–60.

Ramig, R. E., R. R. Allmaras, and R. I. Papendick, 1983. Water conservation: Pacific Northwest, in H. E. Dregne and W. O. Willis, Eds., Dryland Agriculture, Agronomy Monograph 23, American Society of Agronomy, Madison, WI, 105–124.

Rasmussen, P. E. and C. L. Douglas, Jr., 1992. The influence of tillage and cropping-intensity on cereal response to nitrogen, sulfur, and phosphorus, *Fert. Res.*, 31, 15–19.

Rasmussen, P. E. and C. R. Rohde, 1988. Long-term tillage and nitrogen fertilization effects on organic nitrogen and carbon in a semiarid soil, *Soil Sci. Soc. Am. J.*, 52, 1114–1117.

Rasmussen, P. E. and C. R. Rohde, 1989. Soil acidification from ammonium-nitrogen fertilization in moldboard plow and stubble-mulch wheat fallow tillage, *Soil Sci. Soc. Am. J.*, 53, 119–122.

Rasmussen, P. E., H. P. Collins, and R. S. Smiley, 1989. Long-Term Management Effects on Soil Productivity and Crop Yield in Semi-Arid Regions of Eastern Oregon, Station Bulletin 675, Oregon State University Agricultural Experiment Station, 57 pp.

Rickman, R. W., B. Klepper, and R. K. Belford, 1985. Developmental relationships among roots, leaves, and tillers in winter wheat, in W. Day and R. K. Atkins, Eds., *Wheat Growth and Modeling*, Plenum, New York, 83–98.

Roe, R. D. and P. R. Rogers, 1987. Effect of increased crop residue on soil erosion and wildlife habitat potential: intermediate precipitation zone, Whitman County, WA, in L. F. Elliott, Ed., *STEEP—Conservation Concepts and Accomplishments*, Washington State University Press, Pullman, 629–631.

Smiley, R. W., 1992. Estimate of cultivated acres for agronomic zones in the Pacific Northwest, in Columbia Basin Agricultural Research Spec. Rpt. 894, Oregon State University, Corvallis, 86–87.

Smiley, R. W., A. G. Ogg, Jr., and R. J. Cook, 1992. Influence of glyphosate on *Rhizoctonia* root rot, growth, and yield of barley, *Plant Dis.*, 76, 937–942.

Thill, D. C., V. L. Cochran, F. L. Young, and A. G. Ogg, Jr., 1987. Weed management in annual cropping limited-tillage systems, in L. F. Elliott, Ed., *STEEP—Conservation Concepts and Accomplishments*, Washington State University Press, Pullman, 275–287.

U.S. Department of Agriculture, Soil Conservation Service, 1981. Land Resource Regions and Major Land Resource Areas of the United States, Agricultural Handbook 296, USDA-SCS, Washington, D.C.

Veseth, R., K. Saxton, and D. McCool, 1992. Tillage and residue management strategies for variable cropland, in PNW Conservation Handbook Series, chap. 3, Residue Management No. 18. University of Idaho, Moscow, 1–10.

Wiese, M. V., Ed., 1987. *Compendium of Wheat Diseases*, 2nd ed., American Phytopathological Society, St. Paul, MN, 112 pp.

Wiese, M. V., R. J. Cook, D. M. Weller, and T. D. Murray, 1987. Life cycles and incidence of soil borne plant pathogens in conservation tillage systems, in L. F. Elliott, Ed., *STEEP—Conservation Concepts and Accomplishments*, Washington State University Press, Pullman, 299–313.

Wilkins, D. E., 1988. Apparatus for Placement of Fertilizer below Seed with Minimum Soil Disturbance, U.S. Patent 4,765,263, Aug. 23, 1988.

Wilkins, D. E., E. A. Dowding, G. M. Hyde, C. L. Peterson, and G. J. Swanson, 1987. Conservation tillage equipment for seeding, in L. F. Elliott, Ed., *STEEP—Conservation Concepts and Accomplishments*, Washington State University Press, Pullman, 571–575.

Young, R. A., M. J. M. Romkens, and D. K. McCool, 1990. Temporal variations in soil erodibility, in Bryan, R. B., Ed., *Soil Erosion—Experiments and Models*, Catena Suppl. 17, Catena Verlag, New York.

Zuzel, J. F. and J. L. Pikul, Jr., 1987. Infiltration into a seasonally frozen agricultural soil, *J. Soil Water Conserv.*, 42, 447–450.

Zuzel, J. F., J. L. Pikul, Jr., and R. N. Greenwalt, 1986. Point probability distributions of frozen soil, *J. Climate Appl. Meteorol.*, 11, 1681–1686.

6 Integrated Pest Management for Conservation Systems

Alex G. Ogg, Jr., Richard W. Smiley, Keith S. Pike,
Joseph P. McCaffrey, Donn C. Thill, and
Sharron S. Quisenberry

CONTENTS

6.1   INTRODUCTION

Integrated pest management (IPM) can be defined as "the selection, integration and implementation of pest control based on predicted economic, ecological and socio-logical consequences" (Bottrell, 1979). In practical terms, IPM is a decision-making process of managing pests that determines if, when, where, and what strategies or mixture of practices are needed to provide cost-effective, environmentally sound control (Olkowski et al., 1988). Pest management is a component of integrated crop management and it is impossible to discuss IPM without considering the whole production system. In this chapter the use of *pest* will refer to diseases, insects, and weeds.

The commonly accepted definition of conservation tillage is any tillage and planting system that leaves at least 30% of the soil surface covered with residues

0-8493-1185-3/99/$0.00+$.50
© 1999 by CRC Press LLC

after a crop is planted (Unger, 1990). Typically, soils under conservation systems compared with soils under conventional systems are cooler and wetter in the spring and summer and are warmer and wetter in the fall and winter. Thus, conservation systems provide a different habitat than conventional tillage systems for attracting and supporting pests that can attack or interfere with the growth and yield of crops.

As farming practices have changed to reduce water and wind erosion, there have been simultaneous changes in pest species and populations. This chapter briefly outlines the general concepts of IPM, lists and discusses the components of IPM and the changes in pests as farmers adopt conservation farming practices, and describes IPM systems for five agronomic cropping zones in the nonirrigated regions of the Pacific Northwest (PNW).

6.2   GENERAL IPM CONCEPTS

The objective of an IPM program is to control pests economically and to minimize hazards to the environment by coordinating pest management practices with production practices. The key components of IPM are

- Accurate pest identification,
- Field monitoring,
- Control action guidelines,
- Effective pest management methods.

Pests must be identified accurately before the best control strategy can be selected and implemented. Different control measures may be needed for closely related species. If a pest cannot be positively identified, a county agent, industry fieldman, or extension specialist should be contacted before a control measure is selected.

Regular field monitoring is an important practice for good farm management and IPM. For example, noting the date of crop emergence is important for timing of fertilizer applications as well as determining the economic threshold for weeds. Crops that emerge several weeks before weeds are much more competitive and an herbicide application may not be needed. For example, downy brome* that emerges more than 20 days after winter wheat is not very competitive and has minimal impact on wheat grain yield unless weed populations are high (Veseth et al., 1994). Fields should be monitored on a regular basis to note crop growth stages, weed populations and growth stages, appearance of disease symptoms, and development and buildup of insect populations. Accurate information on pests from the previous crop is a good indicator of what and where pests may occur in the current crop. This information is useful especially when herbicides must be applied before weeds emerge.

Control action guidelines are often referred to as economic thresholds. Applying a control measure only when the economic threshold is exceeded is a basic concept

---

* Scientific names of weeds, insects, and diseases mentioned in this chapter are listed in Table 6.14, at the end of the chapter.

of IPM. Guidelines are based usually on field history, crop growth stage, pest populations, and weather. From the grower's point of view, the projected price he or she will receive for the crop also is a factor in determining the economic threshold. Control action guidelines for a given pest (e.g., wild oat) can vary between years, crops, and fields. Environmental conditions and crop vigor greatly affect control action guidelines. For example, a vigorously growing, dense spring barley crop that emerges 1 week before wild oat will be more competitive than barley that emerges with wild oat and will have a higher control action guideline than the later-emerging barley or a slowly growing or sparse crop.

The final component of an IPM program is effective pest management methods. These include such practices as planting certified seed, using crop rotations, selecting competitive cultivars, planting insect- and disease-resistant cultivars when available, selecting and timing tillage operations, placing fertilizer for maximum benefit to the crop, adjusting the timing and rate of seeding, using biological controls when available, and selecting and timing pesticide applications to optimize pest control while minimizing the risk to the environment.

For IPM to be successful, all of the components must be combined carefully into a complete program that also is integrated into the crop production system.

6.3 SHIFTS IN PEST SPECIES AND POPULATIONS

6.3.1 Weeds

The inability to control weeds has limited the adoption of conservation tillage more than any other single factor (Williams and Wicks, 1978; Koskinen and McWhorter, 1986). In conservation tillage systems, weed seeds are concentrated near the soil surface (Wicks and Somerhalder, 1971) and weeds that germinate and establish in surface litter or emerge only from shallow depths are favored in these systems. Weeds that can establish in surface litter include grasses such as downy brome, cheat, jointed goatgrass, wild oat, Italian ryegrass, bulbous blue-grass, and interrupted windgrass. Blackshaw (1994) found that downy brome populations were often higher with zero tillage compared with conventional tillage. Horseweed, panicle willowweed, prickly lettuce, and Canada thistle, which have seeds that are dispersed by wind, are also more troublesome in conservation cropping systems. In addition, perennial weeds such as Canada thistle, field bind-weed, and quackgrass have been more troublesome in conservation tillage systems. In Indiana, Canada thistle became a significant and dominant problem after 15 years of no-till (Hickman, 1995).

As tillage has been reduced to meet conservation compliance, farmers have used more herbicides to control weeds (Thill et al., 1986). Currently, there are no biological agents available commercially for control of common weeds in conservation tillage systems. Also, because of the effectiveness and relatively low cost of herbicides, few efforts have been made to enhance weed suppression with cultural practices. Typically, changing only one cultural practice, such as row spacing or seeding rate, has very little effect on weed competition with dryland crops (Barton et al., 1992; Boerboom and Young, 1995).

In the arid regions of the PNW, Russian thistle, downy brome, jointed goatgrass, and kochia are the most troublesome annual weeds in conservation cropping systems. In the intermediate-rainfall zones, kochia, horseweed, prostrate knotweed, and wild oat are also troublesome. In the high-rainfall zones, the most troublesome weeds in conservation cropping systems include downy brome, jointed goatgrass, wild oat, mayweed chamomile, and catchweed bedstraw. To control these weeds in conservation tillage systems, producers rely heavily on nonselective herbicides applied before planting and in fallow and on selective herbicides in the crop. In arid regions where winter wheat is planted with deep-furrow drills into moist soil in stubble-mulch fallow, early plantings will often give the crop an early competitive advantage over downy brome (Appleby and Morrow, 1990; Veseth et al., 1994). In the annual cropping, higher-rainfall regions, delaying planting of winter wheat until mid-October will allow time for the first flush of downy brome to emerge so that it can be killed with herbicides, thus reducing competition with winter wheat (Appleby and Morrow, 1990). At present, there are no herbicides that will control jointed goatgrass selectively in winter wheat; thus, farmers must avoid growing winter cereals for 3 to 5 years to keep this weed suppressed. In arid regions, Russian thistle is especially troublesome and producers rely on a combination of in-crop and postharvest herbicide treatments to manage this weed in conservation tillage systems (Young et al., 1995).

In 1995, grading standards for soft white wheat became significantly more restrictive in relation to allowed foreign material, including weed seeds. This placed new emphasis on weed control, and unfortunately will encourage producers to use more herbicides to ensure clean grain. Also, the new standards will restrict the adoption of conservation tillage practices when weeds such as jointed goatgrass are present. Jointed goatgrass cannot be controlled selectively in winter wheat, and it is difficult and expensive to remove jointed goatgrass spikelets from the harvested grain.

The relative effectiveness of various management strategies for weeds especially troublesome in conservation cropping systems is shown in Table 6.1 for winter wheat, Table 6.2 for spring barley and spring wheat, Table 6.3 for spring pea and spring lentil, and Table 6.4 for chickpea.

Spring canola and winter rapeseed require finely prepared, firm seedbeds for good plant establishment, and to date these crops have not been adapted to conservation cropping systems; therefore, shifts in weed species because of changing tillage practices are unknown. In conventional tillage systems, weedy mustards are the most troublesome weeds. The most common mustard family weeds found in PNW spring canola fields are wild mustard, black mustard, birdsrape mustard, tumble mustard, field pennycress, flixweed, and shepherd's-purse (Davis et al., 1994). Oil content of seeds of these weeds ranged from 24 to 38%, compared with 40 to 50% for rapeseed (Kephart and Murray, 1990). When canola seed is processed, oil from the weed seeds is also extracted and it will contaminate the canola oil. Erucic acid, an undesirable component in edible canola oil, was particularly high in the oil of birdsrape mustard (47%), field pennycress (37%), and black mustard (34%). Also, the glucosinolate content of all of the weedy mustards exceeded 100 μmol/g, and greatly exceeded the typical amount (20 μmol/g) in canola oil. Because these weedy mustards occur in nearly every field in the PNW and because their seeds are long-lived in the soil, these

TABLE 6.1
Relative Effectiveness of Management Strategies for Control of Weeds Especially Troublesome in Winter Wheat Grown in Conservation Farming Systems[a]

Weeds	Tillage Practice[b]		Certified Seed	Crop Rotation[c]			Seeding		Nitrogen		Herbicides[g]	Postplant	
	No-Till	Minimum		2-yr	3-yr	4-yr+	Date[d]	Rate[e]	Rate[e]	Placement[f]	Preplant	Preemerge	Postemerge
Downy brome (cheatgrass)	-1	-1	1	1	3	3	1E	1	-1A	1	2	1	2
Jointed goatgrass	-1	-1	3	1	1	2	1E	1	-1A	1	1	0	0
Wild oat	0	-1	2	0	1	2	1E	1	-1A	1	2	1	3
Russian thistle	-1	0	1	0	1	2	1E	1	0	1	0	2	3
Kochia	-1	0	1	0	1	2	1E	1	0	1	0	2	3
Prickly lettuce	-1	-1	1	1	2	2	1E	1	0	1	2	2	3
Horseweed	-1	0	1	1	1	2	1E	1	0	1	2	2	3
Catchweed bedstraw	-1	-1	1	0	0	1	0	1	0	1	0	1	3
Prostrate knotweed	-1	-1	1	0	1	2	1E	1	0	1	0	1	3
Canada thistle	-1	-1	2	1	1	1	0	1	0	1	2	0	2

[a] Rating scale of effectiveness: -1 = makes worse; 0 = no effect; 1 = slight benefit; 2 = moderate benefit; 3 = highly effective.
[b] Compared to conventional tillage.
[c] Compared to continuous winter wheat.
[d] L = later better; E = earlier better.
[e] Compared to recommended. B = below recommended; A = above recommended.
[f] Placing nitrogen below the seed at planting compared to broadcasting nitrogen.
[g] Refer to the current year's PNW Weed Control Handbook.

TABLE 6.2
Relative Effectiveness of Management Strategies for Control of Weeds Especially Troublesome in Spring Barley and Spring Wheat Grown in Conservation Farming Systems[a]

| Weeds | Tillage Practice[b] | | Certified Seed | Crop Rotation[c] | | | Seeding | | Nitrogen | | Herbicides[g] | | |
	No-Till	Minimum		2-yr	3-yr	4-yr+	Date[d]	Rate[e]	Rate[e]	Placement[f]	Preplant	Postplant Preemerge	Postemerge
Wild oat	-1	-1	2	1	2	2	1E	1A	-1A	2	2	NA[h]	3
Russian thistle	-1	-1	1	2	3	3	1E	1A	0	1	NA	NA	3
Kochia	-1	-1	1	2	3	3	1E	1A	0	1	NA	NA	3
Prickly lettuce	-1	-1	1	1	2	3	1L	1A	0	1	NA	NA	3
Horseweed	-1	-1	1	0	1	1	1L	1A	0	1	NA	NA	3
Catchweed bedstraw	-1	-1	1	0	1	2	1L	1A	0	1	NA	NA	2
Prostrate knotweed	° -1	-1	1	1	2	2	1L	1A	0	1	NA	NA	2
Canada thistle	-1	-1	2	1	1	1	0	1	0	0	2	0	2

[a] Rating scale of effectiveness: -1 = makes worse; 0 = no effect; 1 = slight benefit; 2 = moderate benefit; 3 = highly effective.
[b] Compared to conventional tillage.
[c] Compared to continuous spring barley or spring wheat.
[d] L = later better; E = earlier better.
[e] Compared to recommended. B = below recommended; A = above recommended.
[f] Placing nitrogen below the seed at planting compared to broadcasting and incorporating.
[g] Refer to the current year's PNW Weed Control Handbook.
[h] NA = none available.

TABLE 6.3
Relative Effectiveness of Management Strategies for Control of Weeds Especially Troublesome in Spring Pea and Lentil Grown in Conservation Farming Systems[a]

Weeds	Tillage Practice[b]		Certified Seed	Crop Rotation[c]		Seeding		Herbicides[f]		
	No-Till	Minimum		3-yr	4-yr+	Date[d]	Rate[e]	Preplant	Postplant Preemerge	Postemerge
Wild oat	0	−1	2	2	2	1E	1A	2	0	3
Shepherd's-purse	0	−1	1	2	2	0	1A	3	2	2
Lambsquarters	1	1	1	2	2	0	1A	2	2	3
Prickly lettuce	−1	0	1	2	2	0	1A	3	2	2
Hairy nightshade	1	0	2	1	2	0	1A	2	2	2
Mayweed chamomile	−1	0	1	1	2	0	1A	2	2	3[g]
Wild buckwheat	−1	−1	1	2	2	0	1A	2	2	2
Canada thistle	−1	−1	2	1	1	0	1A	1	0	1

[a] Rating scale of effectiveness: −1 = makes worse; 0 = no effect; 1 = slight benefit; 2 = moderate benefit; 3 = highly effective.
[b] Compared to conventional tillage.
[c] Compared to a winter wheat chickpea rotation.
[d] L = later better; E = earlier better.
[e] Compared to recommended. B = below recommended; A = above recommended.
[f] Refer to current year's PNW Weed Control Handbook.
[g] Based on use of bentazon. Registered for use in pea but not lentil.

TABLE 6.4
Relative Effectiveness of Management Strategies for Control of Weeds Especially Troublesome in Chickpea Grown in Conservation Farming Systems^a

Weeds	Tillage Practice[b]		Certified Seed	Crop Rotation[c]		Seeding		Herbicides[f]		
	No-Till	Minimum		3-yr	4-yr+	Date[d]	Rate[e]	Preplant	Postplant Preemerge	Postemerge
Wild oat	0	−1	2	1	2	1E	0	2	0	3
Shepherd's-purse	0	−1	1	2	2	0	0	3	2	NA[g]
Lambsquarters	1	1	1	2	2	0	0	2	2	NA
Prickly lettuce	−1	0	1	2	2	0	0	3	2	NA
Nightshade	1	0	2	1	2	0	0	2	2	NA
Mayweed chamomile	−1	0	1	1	2	0	0	2	2	NA
Wild mustard	1	0	1	2	2	0	0	3	2	NA
Wild buckwheat	−1	−1	1	2	2	0	0	2	2	NA
Canada thistle	−1	−1	2	1	1	0	0	1	0	NA

[a] Rating scale of effectiveness: −1 = makes worse; 0 = no effect; 1 = slight benefit; 2 = moderate benefit; 3 = highly effective.
[b] Compared to conventional tillage.
[c] Compared to winter wheat chickpea rotation.
[d] L = later better; E = earlier better.
[e] Compared to recommended. B = below recommended; A = above recommended.
[f] Refer to current year's *PNW Weed Control Handbook.*
[g] NA = No herbicides registered for postemergence application.

weeds will be problems in both conventionally tilled and minimum-tilled fields. Avoiding fields infested heavily with weedy mustards is the only strategy available to manage these weeds in spring canola and winter rapeseed. In canola fields infested with shepherd's-purse or field pennycress, the amount of weed seed harvested with the canola seed can be minimized by raising the cutter bar on the combine as high as possible (Brennan, 1995). The other weedy mustards grow tall, and raising the cutter bar has no benefit in reducing weed seed contamination.

Wild oat is troublesome in spring canola grown in conventionally tilled systems, and undoubtedly this weed will be more troublesome in canola grown in conservation systems. Wild oat increased dramatically in spring pea grown under conservation tillage compared with conventional tillage (Kraft et al., 1991; Young et al., 1996). Wild oat can be controlled selectively in canola with herbicides applied postemergence (Brammer et al., 1995).

Perennial weeds such as Canada thistle, field bindweed, and quackgrass will also be more troublesome in canola and winter rapeseed grown with conservation systems. At present, there are no selective herbicides for controlling Canada thistle and field bindweed in canola. Quackgrass can be controlled selectively in canola with herbicides. Long crop rotations and applying nonselective herbicides during fallow periods are some other strategies for managing perennial weeds in fields to be planted to spring canola and winter rapeseed.

6.3.2 INSECTS

Conservation tillage may increase (All and Musick, 1986), decrease (Troxclair and Boethel, 1984; Zehnder and Linduska, 1987), or have no effect (Funderburk et al., 1990) on pest or beneficial arthropod populations.

The diversity of soil and ground-dwelling arthropods (*Collembola, Acarina, Pscoptera,* and predatory ground-dwelling *Coleoptera*) can be higher in conservation tillage cereals and peas than in these crops grown conventionally (Borden, 1991). Predatory beetles have been shown to increase with conservation tillage systems and may be important biocontrol agents of pest insects (House and Stinner, 1987). Borden (1991) reported that arthropod pest problems were not accentuated by conservation tillage cereals or peas. However, Edwards (1975) reported wireworms and slugs as pests in cereal grown with conservation tillage. Wireworms can also be serious pests of peas and other legumes by feeding on germinating seeds and roots, thus reducing seedling vigor and plant stand. Wireworm populations have the potential to increase in response to decreased tillage (All and Musick, 1986). This has been observed in wheat and corn grown under no-till (Gregory, 1974; Edwards 1975) and fields recently brought out of sod (Musick and Beasley, 1978). However, Williams (1993) found lower wireworm populations in peas grown in conservation cropping systems compared with peas grown in conventional systems. Readers are referred to Robinson et al. (1974) for a complete description and control strategies for insects infesting peas.

Researchers have found increased species diversity of foliage arthropods (House and Stinner, 1983), decreased species diversity (Peterson, 1982; Burton and Krenzer, 1985), or no effect on species diversity (Barney and Pass, 1987; Borden, 1991) in conservation tillage systems compared with conventional tillage systems. Peas supported similar foliage arthropod communities in both conservation and conventional tillage systems (Borden 1991). Borden (1991) found that pea leaf weevil increased in conservation tillage systems, whereas pea aphid decreased. However, Quisenberry and Schotzko (unpublished data, 1995) did not observe differences in pea leaf weevil or pea aphid populations in peas grown in a conservation tillage system compared with conventional systems.

Pea aphids, cowpea aphids, lygus bugs, western yellow-stripped armyworms, and seedcorn maggots are the most economically important insects that attack lentils grown under conventional tillage systems in the Palouse (Muehlbauer et al., 1995). Aphid-vectored pathogenic viruses and direct feeding of aphids can devastate lentils. Lygus bugs feed on immature pods causing seed and pod abortion and a troublesome seed quality problem known as chalky spot (Summerfield et al., 1982). Sometimes large populations of western yellow-stripped armyworm develop late in the growing season and defoliate plants and consume pods. Because lentils are seldom grown under conservation tillage practices in the Palouse, changes in insects or insect populations due to adoption of conservation tillage practices are largely unknown.

Hessian fly, wheat jointworm, and wheat stem sawfly overwinter in straw and, if present, can be more troublesome in wheat grown in conservation tillage systems compared with conventional (Pike et al., 1993; Papendick et al., 1995). Greenbug populations are reduced in wheat grown with moderate to high surface residues (Burton and Krenzer, 1985), whereas English grain aphid populations are increased in winter wheat grown in conservation cropping systems compared with conventional systems (Borden, 1991). Brown wheat mite, wheat curl mite, and winter grain mite on small grains are increased when volunteer grains and weedy grasses are not controlled in summer fallow (Chada, 1956; Atkinson and Grant, 1967; Sloderbeck, 1989a, b).

Although the total number of insects in small grains and legumes is higher in conservation cropping systems compared with conventional systems, the relative importance of the majority of these pests remains similar. There are some exceptions, which are summarized in Tables 6.5 and 6.6.

There are a number of insect pests associated with spring-planted canola (Homan and McCaffrey, 1993) and fall-planted canola and rapeseed (McCaffrey, 1990), and those that are troublesome in PNW are listed in Table 6.7. The effects of conservation tillage systems on insect pest development or the response of the crop to insect damage in the PNW is not well established.

With the exception of cutworms and armyworms, most insect pests associated with spring-planted canola will not be affected directly by tillage systems. However, no-till and conservation tillage may affect indirectly the response of spring-planted canola to damage by the flea beetle. Spring-planted canola requires a finely tilled, firm seedbed to promote rapid germination, uniform emergence, and early stand establishment. Moderate levels of crop residue to reduce erosion are tolerable, but excessive residue levels can reduce germination and emergence of crop seeds (Kephart et al., 1990). Flea beetle damage is a dynamic process, occurring during a short period of time when the plants are small and susceptible to stress (Lamb,

TABLE 6.5

Key Insect Pests of Wheat or Barley Grown under Conservation Farming Systems in the PNW and the Relative Performance of Various Management Strategies

| Insect Pest | Presence | Host Crop | Crop Damage Potential | Main Season for Damage | Tillage Practice[b] | | Biological | Resistant Cultivars | Crop Rotation[c] | | | Seeding Date[d] | | Insecticides[e] | |
					No-Till	Minimum			2-yr	3-yr	4-yr+	Spring	Fall	Soil	Foliar
English grain aphid	Common	b,w[f]	Major[g,h]	Spr/sum	−1	0	1–3	0	0	0	0	2–3E	2–3L	2	2–3
Greenbug	Infrequent	b,w	Major[g,h,i]	Sum/fall	2	1	1–3	0	0	0	0	3E	3L	1	3
Hessian fly	Infrequent[j]	w	Major	Spr/fall	−1	1	0	3	2	3	3	2–3E	3L	3	0
Wheat jointworm	Infrequent	w	Minor	Spr/fall	−1	0	0	0	2	2	2	0	0	0	0
Wheat stem sawfly	Restricted[k]	w	Minor	Spring	−1	1	0	2	3	3	3	0	0	0	0
Wheat strawworm	Infrequent	w	Minor	Spring	−1	−1	0	0	3	2	2	0	0	0	0

a Rating scale of performance: −1 = may make worse; 0 = no effect or not recommended; 1 = slight benefit; 2 = moderate effect; 3 = highly effective.

b Compared to conventional tillage.

c Compared to continuous small grains.

d E = earlier better; L = later better.

e Refer to the current year's *PNW Insect Control Handbook.*

f b = barley; w = wheat.

g Virus vector.

h Uncontrolled green grass or volunteer in summer fallow or stubble increases pest potential.

i Main economic problem centered in southern Idaho during the last 15 years.

j Reaches major pest status only in areas with annual precipitation >20 in.

k Montana pest only, although it is found elsewhere in the region.

TABLE 6.6
Relative Effectiveness of Management Strategies for Control of Insects Especially Troublesome in Spring Legumes Grown in Conservation Farming Systems[a]

Insect	Crop[b]	Tillage Practice[c]		Resistant Cultivars	Crop Rotation[d]		Biological	Seeding Date	Insecticides[f]	
		No-Till	Minimum		3-yr	4-yr+			Soil	Foliar
Seedcorn maggots	P, L, CP	-1	-1	0	NI	NI	0	1L	2-3	0
Wireworms	P, L, CP	0	0	0	3	3	0	0	1-3	0
Thrips	L	-1	-1	0	0	0	0	1E	0	0
Pea leaf weevil	P	-1	-1	0	0	0	1-2	0	0	3
Pea weevil	P	0	0	0	0	0	0	0	0	3
Pea aphid	P, L, CP	1	1	1	0	0	1-3	2E	0	3
Lygus bug	L	-1	-1	0	0	0	0	1E	0	3

[a] Rating scale: -1 = worse; 0 = no effect; 1 = slight benefit; 2 = moderate effect; 3 = highly effective; NI = no information.
[b] Crops abbreviated: P = pea; L = lentil, CP = chickpea.
[c] Compared to conventional tillage.
[d] Compared to legumes in winter wheat/spring legume rotation.
[e] E = earlier; L = later.
[f] Refer to current year's PNW Insect Control Handbook.

TABLE 6.7

Insect Pests of Canola and Rapeseed in the PNW Grown under Conservation Farming Systems, and the Relative Performance of Various Pest Management Strategies[a]

Insect/Mite	Presence	Host[b] Crop	Crop Damage Potential	Main Season for Damage	Tillage No-Till	Tillage Minimum	Biological	Resistant[c] Cultivars	Crop Rotation 2-yr	Crop Rotation 3-yr	Crop Rotation 4-yr+	Seeding Date	Seeding Rate	Insecticides[d,e] Soil	Insecticides[d,e] Foliar
Cabbage flea beetle	Common	sc,fc,fr	Major	Spr	?	?	0	0	0	0	0	1	1	0	1
Cabbage seedpod weevil	Common	fr,fc,sc	Major	Spr/sum	0	0	0	0	0	0	0	1	0	0	3
Aphids															
Cabbage aphid	Common	sc,fc,fr	Major	Sum	0	0	1	0	0	0	0	1	0	0	2
Turnip aphid	Common	sc,fc,fr	Major	Sum	0	0	1	0	0	0	0	1	0	0	2
Green peach aphid	Infrequent	sc	Minor	Sum	0	0	1	0	0	0	0	1	0	0	1
Diamondback moth	Common	sc,fr,fc	Major	Sum/fall	0	0	0	0	0	0	0	1	0	0	3
Lygus bugs	Infrequent	sc,fc,fr	Minor	Sum	0	0	0	0	0	0	0	0	0	0	0
Cutworms	Infrequent	sc,fc,fr	Minor	Spr/sum	?	?	0	0	0	0	0	0	0	0	0
Loopers/armyworms	Infrequent	sc,fc,fr	Minor	Spr/sum	?	?	0	0	0	0	0	0	0	0	0
Painted lady butterfly	Infrequent	sc	Minor	Sum	0	0	0	0	0	0	0	0	0	0	0

[a] Rating scale of performance: –1 = may make worse; 0 = no effect or not recommended; 1 = slight benefit; 2 = moderate effect; 3 = highly effective; ? = effect not known.

[b] sc = spring-planted canola; fc = fall-planted canola; fr = fall-planted rapeseed.

[c] No resistant cultivars available. Research in progress.

[d] 0 rating may reflect the lack of a legally registered insecticide.

[e] Refer to current year's *PNW Insect Control Handbook.*

1984). The ability to grow rapidly in spite of damage (tolerance) is an important mode of resistance to flea beetle attack for canola and rapeseed (Brandt and Lamb, 1994). Environmental conditions that lead to poor stand establishment and seedling growth may interact negatively with the genetically based capacity of the plant to tolerate flea beetle damage. Recently, Milbrath et al. (1995) reported fewer flea beetles per plant in no-till vs. conventional tillage spring-planted canola plots in North Dakota, but plant damage or yield response was not addressed. The interactions of reduced tillage systems and the response of the canola plant to flea beetle damage deserves further study.

Fall-planted canola and rapeseed are planted after summer fallow because of moisture requirements necessary to establish the crop (Kephart and Murray, 1990). Again, a moderate amount of crop residue in the seedbed to reduce erosion is acceptable, but a uniform distribution of crop residue is essential for good seedbed preparation (Kephart and Murray, 1990). Insect pests associated with the seedling and overwintering rosette stage are not usually a problem (McCaffrey, 1990) (see Table 6.7); however, flea beetle damage to seedlings can occur in some areas if the crop is planted too early. High populations of cabbage aphid and diamondback moth can develop on rosettes in the late fall, but this is rare and usually results from extended, warm fall weather that favors their development. The major late-season (bloom to postbloom) pest of fall-planted canola and rapeseed is the cabbage seedpod weevil (McCaffrey, 1990). Conservation farming should have little direct effect on this pest whose larvae feed on seeds developing in pods.

6.3.3 DISEASES

Foliar and root diseases are significant constraints to crop production in the dryland regions of the PNW. As farmers have adopted conservation farming practices, certain diseases have become more troublesome (Cook and Murray, 1986; Wiese et al., 1986).

Fusarium foot rot and *Pythium* root rot are especially severe in continuous winter wheat or in wheat grown in 2-year rotations under no-till or conservation tillage systems (Table 6.8). These diseases are caused by fungi that carry over from year to year in the soil or in crop residues (Cook, 1980; Cook et al., 1980; Veseth et al., 1993). The use of resistant cultivars, seed treatments, 3- to 4-year rotations, or adjusting planting dates can reduce the severity of these diseases. Take-all, caused by a soilborne fungus that lives on roots and crowns of previous wheat crops, is more troublesome in conservation tillage systems, especially no-till (Moore and Cook, 1984). Take-all can be controlled by not growing wheat more than once every 3 years (Cook and Veseth, 1991).

In spring wheat, *Fusarium* foot rot, *Pythium* root rot, *Rhizoctonia* root rot, and take-all occur frequently in conservation cropping systems (Moore and Cook, 1984; Weller et al., 1986) (Table 6.9). *Pythium* root rot can be reduced by later planting and by seed treatment. *Rhizoctonia* root rot can be controlled best by eliminating overwintering volunteer grains and weeds ("green bridge") that allow this fungus to carry over the winter and infest spring-seeded crops (Cook and Veseth, 1991; Smiley et al., 1992). As in winter wheat, take-all is best controlled by planting spring wheat no more than once every 3 years. Refer to Wiese (1987) for descriptions of other diseases affecting wheat.

TABLE 6.8

Relative Effectiveness of Management Strategies for Control of Diseases Especially Troublesome in Winter Wheat Grown in Conservation Farming Systems[a]

Diseases	Tillage Practice[b]		Resistant Cultivars	Crop Rotation[c]			Seeding Date[d]	Fungicides[e]	
	No-Till	Minimum		2-yr	3-yr	4-yr+		Seed	Foliar
Barley yellow dwarf	0	0	0	0	0	0	3L	1	0
Black chaff	−1	−1	0	2	2	2	0	0	0
Cephalosporium stripe	−1	−1	2	1	3	3	3L	0	0
Fusarium foot rot	−1	−1	2	0	1	2	2L	0	0
Physiologic leaf spot	0	0	2	2	3	3	3L	0	0
Powdery mildew	−1	−1	3	2	2	2	2L	1	3
Pythium root rot	−1	−1	0	1	2	2	2E	2	0
Rhizoctonia root rot	−1	0	0	2	2	3	1E	0	0
Septoria leaf blotch	−1	−1	2	2	3	3	2L	1	3
Strawbreaker foot rot	1	0	3	1	2	3	2L	0	3
Stripe rust	0	0	3	0	0	0	2L	1	3
Take-all	−1	0	0	2	3	3	2E	1	0

[a] Rating scale of effectiveness: −1 = makes worse; 0 = no effect; 1 = slight benefit; 2 = moderate benefit; 3 = highly effective.

[b] Compared to conventional tillage.

[c] Compared to continuous winter wheat.

[d] L = later better; E = earlier better.

[e] Refer to current year's *PNW Disease Control Handbook*.

In spring barley, net blotch, *Pythium* root rot, *Rhizoctonia* root rot, scald, and take-all are more troublesome in conservation tillage systems than in conventional tillage systems (Table 6.10). Net blotch can be controlled by crop rotation and seed treatment, whereas scald can be controlled by crop rotation and foliar fungicides. *Rhizoctonia* root rot can be controlled by eliminating the green bridge (Cook and Veseth, 1991; Smiley et al., 1992). Refer to Mathre (1982) for descriptions of other diseases affecting barley.

Ascochyta blight and *Sclerotinia* vine rot are foliar diseases that are more troublesome in peas grown with conservation tillage than with conventional tillage (Table 6.11). Fungi that cause these two diseases carry over from one year to the next on infected vines (Hagedorn, 1984). Disking or plowing after pea harvest buries the residue and prevents dispersal of the spores; however, in conservation tillage systems the infected residue is retained on the soil surface and spores can be dispersed by wind and rain. Growing peas once every 3 years will control these two diseases. Root rot of pea is worse in no-till than conventional tillage, but is less of a problem in conservation tillage than in conventional tillage. Once fields are infested with *Aphanomyces*, *Thielaviopsis*, and *Fusarium*, peas should not be grown for 4 years or longer (Hagedorn, 1984). *Pythium* root rot can be suppressed with a seed treatment and by planting high-vigor seed.

TABLE 6.9

Relative Effectiveness of Management Strategies for Control of Diseases Especially Troublesome in Spring Wheat Grown in Conservation Farming Systems[a]

Diseases	Tillage Practice[b]		Resistant Cultivars	Crop Rotation[c]			Seeding Date[d]	Fungicides[e]	
	No-Till	Minimum		2-yr	3-yr	4-yr+		Seed	Foliar
Barley yellow dwarf	0	0	0	0	0	0	3E	0	0
Black chaff	0	0	0	2	2	2	0	0	0
Fusarium foot rot	−1	0	0	1	2	2	0	0	0
Pythium root rot	−1	−1	0	1	2	2	2L	2	0
Rhizoctonia root rot	−1	−1	0	2	2	3	1L	1	0
Rusts	0	0	3	0	0	0	2E	1	3
Take-all	−1	0	0	2	3	3	2L	2	0

[a] Rating scale of effectiveness: −1 = makes worse; 0 = no effect; 1 = slight benefit; 2 = moderate benefit; 3 = highly effective.
[b] Compared to conventional tillage.
[c] Compared to continuous spring wheat.
[d] L = later better; E = earlier better.
[e] Refer to current year's *PNW Disease Control Handbook*.

TABLE 6.10

Relative Effectiveness of Management Strategies for Control of Diseases Especially Troublesome in Spring Barley Grown in Conservation Farming Systems[a]

Diseases	Tillage Practice[b]		Resistant Cultivars	Crop Rotation[c]			Seeding Date[d]	Fungicides[e]	
	No-Till	Minimum		2-yr	3-yr	4-yr+		Seed	Foliar
Barley yellow dwarf	0	0	0	0	0	0	2E	0	0
Net blotch	−1	−1	3	1	3	3	0	2	0
Pythium root rot	−1	−1	0	1	2	2	2L	2	0
Rhizoctonia root rot	−1	−1	0	2	2	3	1L	1	0
Rusts	0	0	3	0	0	0	2E	1	3
Scald	−1	−1	2	1	3	3	0	0	3
Take-all	−1	0	0	2	3	3	2L	2	0

[a] Rating scale of effectiveness: −1 = makes worse; 0 = no effect; 1 = slight benefit; 2 = moderate benefit; 3 = highly effective.
[b] Compared to conventional tillage.
[c] Compared to continuous spring barley.
[d] L = later better; E = earlier better.
[e] Refer to current year's *PNW Disease Control Handbook*.

TABLE 6.11

Relative Effectiveness of Management Strategies for Control of Diseases Especially Troublesome in Pea Grown in Conservation Farming Systems[a]

Diseases	Tillage Practice[b]		Resistant Cultivars	Crop Rotation[c]		Seeding Date[d]	Fungicides[e]	
	No-Till	Minimum		3-yr	4-yr+		Seed	Foliar
Foliar								
Ascochyta blight	−1	−1	0	1	2	1L	2	3
Powdery mildew	0	0	0	1	2	1E	0	3
Scerotinia vine rot	−1	−1	0	2	3	0	0	0
Viruses	1	1	0	0	0	1E	0	0
Root								
Fusarium wilt								
Race 1	1	1	3	1	2	0	0	0
Race 2	1	1	0	1	2	0	0	0
Root rot complex	−1	2	0	2	3	1E	1	1
(*Aphanomyces,*								
Fusarium,								
Pythium,								
Rhizoctonia, and								
Thielaviopsis)								

[a] Rating scale of effectiveness: −1 = makes worse; 0 = no effect; 1 = slight benefit; 2 = moderate benefit; 3 = highly effective.
[b] Compared to conventional tillage.
[c] Compared to winter wheat-pea rotation.
[d] L = later better; E = earlier better.
[e] Refer to current year's *PNW Disease Control Handbook*.

Ascochyta blight has been a major disease problem in chickpea, and its severity has been increased by conservation farming systems (Table 6.12). Recently, two new cultivars of chickpea, 'Dwelly' and 'Sanford,' have been released that are tolerant to *Ascochyta* blight (Wiese et al., 1995). At present, there are no cultivars of lentil that are resistant to *Ascochyta* blight. Foliar sprays will control *Ascochyta* blight, but must be applied several times. Also, long crop rotations between legume crops will suppress *Ascochyta* blight. Seedling damping-off of chickpea and lentil is worse in conservation tillage systems than in conventional tillage systems. This disease can be controlled with long rotations (3 years or more between legume crops) and with seed treatments.

Blackleg and *Sclerotinia* stem rot of rapeseed are more problematic in conservation farming systems than in conventional tillage systems (Table 6.13). Blackleg can be controlled by planting resistant cultivars or with seed treatment, and *Sclerotinia* stem rot can be suppressed with foliar fungicides. Both diseases can also be controlled with long rotations where rapeseed is not grown for 3 years after legumes or another cruriferous crop (Kephart and Murray, 1990).

TABLE 6.12
Relative Effectiveness of Management Strategies for Control of Diseases Especially Troublesome in Lentil and Chickpea Grown in Conservation Farming Systems[a]

Diseases	Tillage Practice[b]		Resistant Cultivars	Crop Rotation[c]		Seeding Date[d]	Fungicides[e]	
	No-Till	Minimum		3-yr	4-yr+		Seed	Foliar
Ascochyta blight	−1	−1	2	3	3	1L	2	3
Seedling damping-off	−1	−1	0	0	0	2	3	0
Viruses	1	1	0	0	0	1E	0	0

[a] Rating scale of effectiveness: −1 = makes worse; 0 = no effect; 1 = slight benefit; 2 = moderate benefit; 3 = highly effective.
[b] Compared to conventional tillage.
[c] Compared to winter wheat/lentil/chickpea rotation.
[d] L = later better; E = earlier better.
[e] Refer to current year's *PNW Disease Control Handbook*.

TABLE 6.13
Relative Effectiveness of Management Strategies for Control of Diseases Especially Troublesome in Rapeseed/Canola Grown in Conservation Farming Systems[a]

Diseases	Tillage Practice[b]		Resistant Cultivars	Crop Rotation[c]		Seeding Date[d]	Fungicides[e]	
	No-Till	Minimum		3-yr	4-yr+		Seed	Foliar
Blackleg	−1	−1	3	2	3	0	3	0
Sclerotinia stem rot	−1	−1	0	2	3	0	0	2

[a] Rating scale of effectiveness: −1 = makes worse; 0 = no effect; 1 = slight benefit; 2 = moderate benefit; 3 = highly effective.
[b] Compared to conventional tillage.
[c] Compared to winter wheat/rapeseed/canola rotation.
[d] L = later better; E = earlier better.
[e] Refer to current year's *PNW Disease Control Handbook*.

6.4   IPM PROGRAMS BY AGRONOMIC ZONES

This section describes anticipated pest problems for conservation cropping systems in selected agronomic zones in the PNW and lists IPM strategies to control these pests, to maintain long-term economic viability, and to minimize environmental risks. Detailed descriptions of individual agronomic zones in the PNW are given in Chapter 5 (Douglas et al., this publication) and by Douglas et al., (1990).

6.4.1   THE MOUNTAIN ZONE—ZONE 1

This zone is cold and moist with steep slopes or mountain valleys and short growing seasons. There is no single dominant crop rotation in this zone. In Oregon, winter wheat/spring barley, winter wheat/spring barley/fallow and winter wheat/fallow rotations are used. Winter wheat/spring barley or spring wheat and winter wheat/spring pea are typical rotations in Idaho. In south-central Washington, growers in Zone 1 use a winter wheat/spring pea/winter wheat/spring cereal rotation or they may plant winter wheat for several consecutive years before including spring barley. Also, bluegrass for seed (5 to 7 years) followed by two cycles of spring lentil/winter wheat is used in Zone 1 in Idaho and Washington. This rotation has excellent soil conservation benefits and will be examined in detail for IPM strategies. Typically, lentil is planted no-till after bluegrass and winter wheat is planted no-till after lentil. Because bluegrass requires a finely tilled seedbed for good emergence, it is planted in the spring into conventionally tilled fields.

Pest Type	Pest	IPM Strategies
		Bluegrass (5 years)
Weeds	Wild oat	Wild oat is very competitive with seedling bluegrass. Plant bluegrass at recommended seeding rate; in seedling bluegrass, spray postemergence with primisulfuron (Beacon™) or use weed-wiper to apply glyphosate (Roundup RT™) when wild oat becomes taller than bluegrass; 10 days later mow to cut wild oat heads; repeat wiping and mowing if necessary.
	Downy brome	Plant spring crops for 2 years to reduce interrupted weed
	Interrupted windgrass	seed in the soil. In seedling bluegrass, spot-spray postemergence with primisulfuron as needed. In established stands use pre- or postemergence herbicides.
	Quackgrass	Quackgrass is a prohibited noxious weed in bluegrass seed. Do not plant bluegrass in fields with known infestations of quackgrass. If patches of quackgrass appear, spot-spray with glyphosate or use weed-wipers to apply glyphosate when quackgrass is taller than bluegrass. In seedling bluegrass, spray postemergence with primisulfuron.
	Field pennycress	Field pennycress (fanweed) is a prohibited noxious weed
	Pigweeds	in bluegrass seed shipped common into some states.
	Common lambsquarters	Broadleaf annual weeds are very competitive with
	Pineappleweed	seedling bluegrass, and new seedings almost always
	Mayweed chamomile	need to be sprayed to prevent stand loss. Apply
	Catchweed bedstraw	herbicides postemergence when weeds are less than 1 in.
	Mustards	in diameter. Use pre- or postemergence herbicides in
	Henbit	established stands.
Insects	None	There are no major insect problems in bluegrass grown under nonirrigated conditions, especially when bluegrass is grown for only 5 years.

Pest Type	Pest	IPM Strategies
Diseases	Rust	Plant resistant cultivar; apply fungicides when rust appears on flag leaf, upper stem, or seed head.
	Powdery mildew	Plant resistant cultivar; apply fungicides when powdery mildew appears on 15 to 20% of the leaves.

Spring Lentil

Weeds	Bluegrass	Apply glyphosate in the spring to kill bluegrass.
	Wild oat	Use high seeding rate and narrow rows; monitor weed populations and apply postemergence herbicides when wild oat exceeds 1 plant/yd^2.
	Volunteer blugrass	Spot-spray with postemergence grass herbicide.
	Mayweed chamomile	Broadleaf weeds are very competitive with prickly lettuce
	Prickly lettuce	lentils. Use high seeding rate and narrow rows; apply
	Mustards	herbicides postplant preemergence. If weed populations exceed 2 to 4/yd^2, apply metribuzin (Lexone™/Sencor™) postemergence.
Insects	Seedcorn maggot	Delay seeding until soil is warm so that crop seeds germinate and emerge rapidly; treat seed with insecticide.
	Lygus bug	Early planting will promote early seed development and will reduce damage from lygus bug; monitor populations and apply foliar insecticides when lygus bug populations exceed 7 to 10 per 25 sweeps.
	Pea aphid	Plant lentils as early as possible; monitor populations and apply foliar insecticides when aphids exceed 30 to 40 per one 180° sweep and few beneficials are present. If beneficials are present, wait 1 to 2 days and sample again. Insecticide application may not be necessary if aphid populations remain the same or decrease. Treat the field immediately if aphid populations have increased.
Diseases	*Ascochyta* blight	Extend crop rotation so lentils are grown only once every 4 years or longer. Foliar fungicides will control *Ascochyta* blight, but several applications may be required.
	Damping-off	Delay seeding until soil is warm; treat seed with fungicide.

Winter Wheat

Weeds	Downy brome	Plant tall, fast-growing varieties; plant early and use high seeding rate; band nitrogen below seed at planting; apply metribuzin postemergence when downy brome populations exceed 3 to 5 plants/ft^2; use the lower threshold number when downy brome emerges within 7–10 days after the wheat.
	Wild oat	Plant tall, fast-growing varieties; plant early and use high seeding rate; band nitrogen below seed at planting. Apply herbicide postemergence if wild oat populations exceed 3 to 5 plants/ft^2.

Pest Type	Pest	IPM Strategies
	Prickly lettuce	Plant early and use high seeding rate; band nitrogen below
	Mayweed chamomile	seed at planting. Apply herbicides postemergence when
	Catchweed bedstraw	combined broadleaf weed populations exceed 1/ft^2.
	Mustard spp.	
Insects	None	Insects rarely reach threshold levels on winter wheat grown in the Zone 1.
Diseases	*Cephalosporium* stripe	Delay planting; plant disease-tolerant variety.
	Snow mold	Plant resistant varieties.

6.4.2   THE PALOUSE ZONE—ZONE 2

This zone is characterized as cool and moist with moderate to steep slopes and high crop productivity. Annual cropping with rotations of winter wheat, spring cereals, and legumes is common. A typical rotation would be winter wheat/spring barley/spring dry pea. In conservation systems, spring crops are grown with reduced tillage (fall chisel-plow), whereas winter wheat is planted no-till after peas.

Pest Type	Pest	IPM Strategies
		Winter Wheat
Weeds	Wild oat	Plant taller wheat cultivar; plant early and use high seeding rate; band nitrogen below seed at planting. When wild oat populations exceed 3 to 5 plants/ft^2, apply herbicide postemergence.
	Catchweed bedstraw	Plant early and use high seeding rate; band nitrogen below
	Prickly lettuce	seed at planting. Apply herbicides postemergence when
	Mustard spp.	combined broadleaf weed populations exceed 1/ft^2.
	Mayweed chamomile	
Insects	None	Insects rarely reach threshold levels on winter wheat grown in the Palouse Zone.
Diseases	Strawbreaker foot rot	Plant resistant cultivar or apply foliar fungicide, if more than 10% of the tillers have stem lesions.
		Spring Barley
Weeds	Wild oat	Plant early and use high seeding rate; band nitrogen below seed at planting. If wild oat population exceed 3 to 5 plants/ft^2 within 1 week after barley emerges, apply herbicides postemergence.
	Prickly lettuce	Use high seeding rate; band nitrogen below seed at
	Mayweed chamomile	planting; apply herbicides postemergence when
	Common lambsquarters	broadleaf weed population exceeds 1/ft^2.
	Mustard spp.	
	Pigweed spp.	
Insects	English grain aphid	Monitor aphid and beneficial insect populations; apply insecticides when aphids exceed 2 to 10 per tiller before the dough stage.

Pest Type	Pest	IPM Strategies
Diseases	*Rhizoctonia* root rot	Eliminate the green bridge (volunteer grains and weeds) by applying nonselective herbicides at least 2 weeks before planting.

Spring Pea

Pest Type	Pest	IPM Strategies
Weeds	Wild oat	If wild oat populations were moderate to high in the previous spring barley crop, apply triallate (Far-Go™) preplant. If triallate is not applied preplant, plant early and use high seeding rate; monitor weed populations and apply postemergence herbicides when wild oat population exceed 1 to 2/yd².
	Mayweed chamomile Shepherd's-purse Wild buckwheat	Use high seeding rate; apply herbicides preplant or postplant preemergence. For mayweed chamomile only, apply bentazon (Basagran™) postemergence when weed population exceeds 1/ft².
	Canada thistle Quackgrass	Spot-spray with glyphosate.
Insects	Pea leaf weevil	Monitor weevil and beneficial insect populations and damage to pea leaves; apply insecticide when leaf area loss exceeds 25% or when there is damage to growing point.
	Pea aphid	Plant peas as early as possible; monitor populations and apply foliar insecticides when aphids exceed 30 to 40 per one 180° sweep and few beneficials are present. If aphids are below the threshold and beneficials are plentiful, wait 1 to 2 days and sample again. Insecticide application may not be necessary if aphid populations remain the same or decrease. Treat the field immediately if aphid populations have increased.
Diseases	Seed rot	Plant seed treated with fungicide.

6.4.3 WHEAT–PEA ZONE—ZONE 3

Zone 3 is characterized as cool, moderately dry with deep soils. Winter wheat/green pea is a common rotation. In conservation systems, winter wheat is planted no-till and green pea is planted in fields that are chisel-plowed in the fall and disked and field cultivated twice in the spring. Fields are usually harrowed after the pea crop is planted. Excessive straw after winter wheat can be reduced by disking shallowly about 2 weeks after wheat harvest.

Pest Type	Pest	IPM Strategies

Winter Wheat

Pest Type	Pest	IPM Strategies
Weeds	Downy brome	If downy brome emerges before wheat is planted, apply glyphosate. Plant taller varieties; plant early and use high seeding rate; band nitrogen below seed at planting; apply metribuzin postemergence when downy brome populations exceed 2 to 3 plants/ft².

Pest Type	Pest	IPM Strategies
	Wild oat	Plant early and use high seeding rate; band nitrogen below seed at planting; apply herbicides where wild oat populations exceed 2 to 3/ft².
	Catchweed bedstraw Prickly lettuce Prostrate knotweed Mustard spp.	Plant early and use high seeding rate; band nitrogen below seed at planting; apply herbicides when combined weed populations exceed 1/ft².
Insects	Russian wheat aphid Bird cherry-oat aphid	Delay planting, use a systemic seed treatment, monitor fields for aphid buildup, and treat with a foliar applied insecticide if 5 to 10% of the plants become infested.
Diseases	*Cephalosporium* stripe	Apply lime to soils with pH less than 5.5; delay planting, plant tolerant variety.
	Fusarium foot rot	Avoid stressing plants with excessive nitrogen fertilizer, delay planting, plant tolerant variety.
	Leaf rust	Plant early, apply foliar fungicide.
	Stripe rust	Plant resistant cultivar, treat seed with fungicide.

Spring Green Pea

Pest Type	Pest	IPM Strategies
Weeds	Wild oat	If field has a history of wild oat, apply and incorporate triallate with last field cultivation, or apply postemergence herbicide if wild oat exceeds 2 to 3/yd².
	Shepherd's-purse Cutleaf nightshade Common lambsquarters	Use high seeding rate and narrow rows; apply herbicides preplant or postplant preemergence. Nightshade berries are prohibited in green peas, hand rogue scattered plants or apply herbicides postemergence. In fields with a known history of nightshades, apply imazethapyr (Pursuit™) preplant and incorporate with last field cultivation.
Insects	Pea leaf weevil	Monitor weevil and beneficial insect populations and damage to pea leaves; apply insecticide when leaf area loss exceeds 25% or when there is damage to growing points.
	Pea aphid	Plant peas as early as possible; monitor aphid populations, apply insecticides when populations reach 30 to 40 aphids per 180° sweep and few, if any, natural enemies are present.
	Cornseed maggot	Plant seed treated with insecticide.
Diseases	*Sclerotinia* vine rot	No management in this rotation and conservation tillage system; pea should not be planted in fields with recent history of this disease.
	Pythium root rot	Plant after soil warms to 50°F, treat seed with fungicide, plant high-vigor seed.

6.4.4 Arid Zone (Shallow Soils)—Zone 4

This zone is characterized as cool and dry with shallow and gently to steeply sloping soils. Annual cropping is common on these shallow soils and winter wheat/spring barley is a typical rotation. In conservation systems, winter wheat and spring barley are planted no-till or into reduced tillage systems.

Pest Type	Pest	IPM Strategies
		Winter Wheat
Weeds	Downy brome	If downy brome emerges before wheat is planted, apply glyphosate. Plant early and use recommended seeding rate. Deep-band nitrogen at planting; avoid excess nitrogen. Apply metribuzin postemergence when downy brome populations exceed 3 plants/ft^2 and downy brome emerges within 14 days after the wheat.
	Russian thistle	Plant early and use recommended seeding rate. Deep-band nitrogen at planting. Apply herbicides postemergence when Russian thistle exceeds 3 to 5 plants/yd^2. After harvest, use shallow tillage with sweeps or use nonselective herbicides to prevent seed production by Russian thistle.
Insects	Russian wheat aphid Bird cherry-oat aphid	Delay planting, use a systemic seed treatment, monitor fields for aphid buildup, and treat with a foliar applied insecticide if 5 to 10% of the plants become infested.
Diseases	Take-all	Delay planting, treat seed with fungicide.
	Fusarium foot rot	Delay planting, plant tolerant variety, avoid stressing plants with excessive nitrogen.
		Spring Barley
Weeds	Wild oat	Plant early and use high seeding rate; band nitrogen below seed at planting, apply herbicides when wild oat populations exceed 3 to 4 plants/ft^2.
	Blue mustard Coast fiddleneck Flixweed Russian thistle Pigweed spp. Common lambsquarters Mustard spp. Kochia	Plant early and use high seeding rate; band nitrogen below seed at planting. Apply herbicides postemergence when combined weed populations exceed 1/ft^2.
Insects	None	Insects rarely reach threshold levels on spring barley grown in this zone.
Diseases	*Rhizoctonia* root rot	Eliminate green bridge by applying glyphosate 2 to 3 weeks before planting, treat seed with fungicide.
	Net Blotch	Treat seed with fungicide.

6.4.5 ARID ZONE (DEEP SOILS)—ZONE 5

Zone 5 is characterized as cool and dry with deep and gently to moderately sloping soils. This zone does not receive sufficient rainfall to allow annual cropping. Nearly all of the cropland is in a winter wheat/fallow rotation. In conservation systems as in conventional systems, wheat is planted with deep furrow drills.

Pest Type	Pest	IPM Strategies
		Winter Wheat
Weeds	Downy brome	Shank-in nitrogen before June 1 in the fallow season; avoid excess nitrogen. Prevent downy brome seed production during fallow. Plant early and use recommended seeding rate. Apply metribuzin postemergence when downy brome populations exceed 3 plants/ft^2 and downy brome emerges within 14 days after the wheat. If a dense infestation occurs, switch to a fallow/spring wheat/fallow rotation before planting winter wheat again.
Insects	Russian wheat aphid	Delay planting, use a systemic seed treatment, monitor fields for aphid buildup, and treat with a foliar applied insecticide if 5 to 10% of the plants become infested.
	Bird cherry-oat aphid	
Diseases	*Fusarium* foot rot	Plant tolerant variety; avoid stressing with excessive nitrogen, delay planting.
	Barley yellow dwarf	Eliminate the green bridge by applying glyphosate 2 to 4 weeks before planting, especially in fields near irrigated crops; treat seed with insecticide to control aphids carrying the virus.
	Stripe rust	Plant resistant variety, treat seed with fungicide.

All agricultural chemicals recommended for use in this report have been registered by the Environmental Protection Agency. They should be applied in accordance with directions on the manufacturer's label as registered under the Federal Insecticide, Fungicide and Rodenticide Act.

TABLE 6.14

Common and Scientific Names of Weeds, Insects, and Diseases Mentioned in this Chapter

Host Crop[a]	Common Name	Scientific Name
		Weeds
SC, FC, FR	birdsrape mustard	*Brassica rapa* L.
SC, FC, FR	black mustard	*Brassica nigra* (L.) W.J.D. Koch
L	bluegrass	*Poa pratensis* L.
BG	blue mustard	*Chorispora tenella* (Pallas) DC.
WW	bulbous bluegrass	*Poa bulbosa* L.
WW, SW, SB, L, P, CP	Canada thistle	*Cirisum arvense* (L.) Scop.
WW, SW, SB	catchweed bedstraw	*Galium aparine* L.
WW	cheat	*Bromus secalinus* L.
SB	coast fiddleneck	*Amsinckia intermedia* Fisch. & May
L, P, CP	common lambsquarters	*Chenopodium album* L.
P	cutleaf nightshade	*Solanum triflorum* Nutt.
WW	downy brome	*Bromus tectorum* L.

TABLE 6.14
Common and Scientific Names of Weeds, Insects, and Diseases Mentioned in this Chapter (continued)

Host Crop[a]	Common Name	Scientific Name
WW, SW, SB	field bindweed	*Convolvulus arvensis* L.
SC, FC, FR	field pennycress	*Thlaspi arvense* L.
SC, FC, FR	flixweed	*Descurainia sophia* (L.) Webb. ex Prahtl
L, P, CP	hairy nightshade	*Solanum sarrachoides* Sendt.
BG	henbit	*Lamium amplexicaule* L.
WW	horseweed	*Conyza canadensis* (L.) Cronq.
WW	interrupted windgrass	*Apera interrupta* (L.) Beauv.
WW	Italian ryegrass	*Lolium multiflorum* Lam.
WW	jointed goatgrass	*Aegilops cylindrica* Host
WW, SW, SB	kochia	*Kochia scoparia* (L.) Schrad.
WW, SW, SB L, P, CP	mayweed chamomile	*Anthemis cotula* L.
WW	panicle willowweed	*Eiplobium paniculatum* Nutt. ex T. & G.
BG	pigweeds	*Amaranthus* spp.
BG	pineappleweed	*Matricaria matricariodes* (Less.) C. L. Porter
WW, SB, L, P, CP	prickly lettuce	*Lactuca serriola* L.
WW, SW, SB	prostrate knotweed	*Polygonum aviculare* L.
BG	quackgrass	*Elytrigia repens* (L.) Nevski
WW, SW, SB	Russian thistle	*Salsola iberica* Sennen & Pau
SC, FC, FR, L, P, CP	shepherd's-purse	*Capsella bursa-pastoris* (L.) Medicus
SC, FC, FR	tumble mustard	*Sysimbrium altissimum* L.
L, P, CP	wild buckwheat	*Polygonum convolvulus* L.
SC, FC, FR, L, P, CP	wild mustard	*Brassica kaber* (DC.) L.C. Wheeler
WW, SW, SB, L, P, CP	wild oat	*Avena fatua* L.
	Insects	
C	armyworm	*Pseudaletia unipuncfa* (Haworth)
C	bertha armyworm	*Mamestra configurata* Walker
WW	bird cherry-oat aphid	*Rhapalosiphum padi* (Linnaeus)
WW	brown wheat mite	*Petrobia latens* (Muller)
C	cabbage aphid	*Brevicoryne brassicae* (Linnaeus)
SC, FC, FR	cabbage flea beetle	*Phyllotreta cruciferae* (Goeze)
C	cabbage seedpod weevil	*Ceutorhynchus assimilis* (Paykull)
L	cowpea aphid	*Aphis craccivora* Koch
C	cutworms	*Agrotis* spp.
C	diamondback moth	*Plutella xylostella* (Linnaeus)
WW, SW	English grain aphid	*Sitobion avenae* (Fabricius)
WW	greenbug	*Schizaphis graminum* (Rondani)
SC	green peach aphid	*Myzus persicae* (Sulzer)

TABLE 6.14
Common and Scientific Names of Weeds, Insects, and Diseases Mentioned in this Chapter (continued)

Host Crop[a]	Common Name	Scientific Name
WW	Hessian fly	*Mayetiola destructor* (Say)
L	lygus bug	*Lygus* spp.
SC	painted lady butterfly	*Vanessa cardui* (Linnaeus)
P,L	pea aphid	*Acyrthosiphon pisum* (Harris)
P	pea leaf weevil	*Sitonia lineatus* (Linnaeus)
P	pea weevil	*Bruchus pisorum* (Linnaeus)
WW, SW	Russian wheat aphid	*Diuraphis noxia* (Mordvilko)
L	seedcorn maggot	*Delia platura* (Meigen)
P	slugs	*Deroceras* spp.
L	thrips	
SC, FC, FR	turnip aphid	*Lipaphis erysimi* (Kaltenbach)
L	western yellow-stripped armyworm	*Spodoptera praefica* (Grote)
WW	wheat curl mite	*Eriophyes tulipae* Kiefer
WW	wheat jointworm	*Tetramesa tritici* (Fitch)
WW	wheat stem sawfly	*Cephus cinctus* Norton
WW	wheat strawworm	*Tetramesa grandis* (Riley)
WW	winter grain mite	*Penthaleus major* (Duges)
P,L	wireworms	*Limonius* spp. and *Ctenicera* spp.
		Diseases
P	*Ascochyta* blight	*Ascochyta pisi* Lib., *Mycosphaerella pinodes* (Berk. & Bloxam) Vestergr., and *Phoma medicaginis* Malbr. & Roum. in Roum.
CP		*Mycosphaerella rabiei* Kovachevski
L		*Ascochyta lentis* Vassiljevsky
WW	barley yellow dwarf	barley yellow dwarf virus
WW	black chaff	*Xanthomonus campestris* pv. *translucens* (Jones et al.) Dye
SC, FC, FR	blackleg	*Phoma lingam* (Tode:Fr.) Desmaz.
WW, SW	Cephalosporium stripe	*Cephalosporium gramineum* Nishikado & Ikata in Nishikado et al.
L, CP	damping-off	*Pythium* Pringsh. spp.
WW, SW	*Fusarium* foot rot	*Fusarium culmorum* (Wm. G. Sm.) Sacc. and *F. graminearum* Schwabe
P	*Fusarium* wilt	*Fusarium oxysporum* Schlechtend.:Fr. f. sp. *pisi* (J. C. Hall) W. C. Snyder & Hanna
SB	net blotch	*Pyrenophora teres* Drechs.
P	pea root rot	*Aphanomyces euteiches* Drechs. f. sp. *pisi* W. F. Pfender & D. J. Hagedorn, *Thielaviopsis basicola* (Berk. & Broome) Ferraris, *Fusarium solani* (Mart.) Sacc. f. sp. *pisi* (F. R. Jones) W. C. Snyder & H. N. Hans.

TABLE 6.14
Common and Scientific Names of Weeds, Insects, and Diseases Mentioned in this Chapter (continued)

Host Crop[a]	Common Name	Scientific Name
WW	physiological leafspot	Unknown (possibly chloride deficiency)
WW	powdery mildew	*Erysiphe graminis* DC.
P		*Erysiphe pisi* DC.
WW, P	*Pythium* root rot	*Pythium* Pringsh. spp.
WW, SW, SB	*Rhizoctonia* root rot	*Rhizoctonia solani* Kuhn and *R. oryzae* Ryker & Gooch
SW, SB	rusts	*Puccinia* Pers.:Pers. spp.
SB	scald	*Rhynchosporium secalis* (Oudem.) J. J. Davis
SC, FC, FR	*Sclerotinia* stem rot	*Sclerotinia sclerotiorum* (Lib.) deBary
P	*Sclerotina* vine rot	*Sclerotinia sclerotiorum* (Lib.) deBary
WW	Septoria leaf blotch	*Septoria tritici* Roberge in Desmaz.
WW	strawbreaker foot rot	*Pseudocercosporella herpotrichoides* (Fron) Deighton
WW	stripe rust	*Puccinia striiformis* Westend.
WW, SW	take-all	*Gaeumannomyces graminis* (Sacc.) Arx & D. Olivier var. *tritici* J. Walker
P, CP, L	viruses	Numerous

[a] Crops designated by letters as follows: BG = bluegrass; C = canola; CP = chickpea; FC = fall canola; FR = fall rapeseed; L = lentil; P = pea; SB = spring barley; SC = spring canola; SW = spring wheat; WW = winter wheat.

REFERENCES

All, J. N. and G. J. Musick, 1986. Management of vertebrate and invertebrate pests, in M. A. Sprague and G. B. Triplett, Eds., *No-Tillage and Surface-Tillage Agriculture: The Tillage Revolution*, John Wiley & Sons, New York, 347–387.

Appleby, A. P. and L. A. Morrow, 1990. The Pacific Northwest, in W. W. Donald, Ed., *Systems of Weed Control in Wheat in North America*, Weed Science Society of America, Champaign, IL, 200–232.

Atkinson, T. G. and M. N. Grant, 1967. An evaluation of streak mosaic losses in winter wheat, *Phytopathology*, 57, 188–192.

Barney, R. J. and B. C. Pass, 1987. Influence of no-tillage on foliage-inhibiting arthropods of alfalfa in Kentucky, *J. Econ. Entomol.*, 80, 1288–1290.

Barton, D. L., D. C. Thill, and B. Shafii, 1992. Integrated wild oat (*Avena fatua*) management affects spring barley (*Hordeum vulgare*) yield and economics, *Weed Technol.*, 6, 129–135.

Blackshaw, R. E., 1994. Rotation affects downy brome (*Bromus tectorum*) in winter wheat (*Triticum aestivum*), *Weed Technol.*, 8, 728–732.

Boerboom, C. M. and F. L. Young, 1995. Effect of postplant tillage and crop density on broadleaf weed control in dry pea (*Pisum sativum*) and lentil (*Lens culinaris*), *Weed Technol.*, 9, 99–106.

Borden, E. E. R., 1991. Community Analysis of Arthropods in Small Grain and Legume Crops Produced under Conservation and Conventional Tillage, Ph.D. dissertation, Washington State University, Pullman, 117 pp.

Bottrell, D. R., 1979. Integrated Pest Management. Council on Environmental Quality, U.S. Government Printing Office, Washington, D.C., 120 pp.

Brammer, T. A., J. S. Brennan, E. J. Bechinski, and D. C. Thill, 1995. Bioeconomic model for grass weed control in spring-planted canola, *West. Soc. Weed Sci. Res. Prog. Rep.*, p. 71.

Brandt, R. N. and R. J. Lamb, 1994. Importance of tolerance and growth rate in the resistance of oilseed rapes and mustards to flea beetles, *Phyllotreta* cruciferae (Goeze) (Coleoptera: Chrysomelidae), *Can. J. Plant Sci.*, 74, 169–176.

Brennan, J. S., 1995. Assessment of Weed Competition in Spring-Planted Canola (*Brassica napus* L.), Ph.D. dissertation, University of Idaho, Moscow, 78 pp.

Burton, R. L. and E. G. Krenzer, Jr., 1985. Reduction of greenbug (Homoptera: Aphididae) populations by surface residues in wheat tillage studies, *J. Econ. Entomol.*, 78, 390–394.

Chada, H. L., 1956. Biology of the winter grain mite and its control in small grains, *J. Econ. Entomol.*, 49, 515–520.

Cook, R. J., 1980. Fusarium foot rot of wheat and its control in the Pacific Northwest, *Plant Dis.*, 64, 1061–1065.

Cook, R. J. and T. D. Murray, 1986. Management of soilborne pathogens of wheat in soil conservation systems, in F. L. Elliott, Ed., *STEEP—Conservation Concepts and Accomplishments*, Washington State University, Pullman, 314–323.

Cook, R. J. and R. J. Veseth, 1991. *Wheat Health Management*, APS Press, St. Paul, MN, 152 pp.

Cook, R. J., J. W. Sitton, and J. T. Waldher, 1980. Evidence for *Pythium* as a pathogen of direct-drilled wheat in the Pacific Northwest, *Plant Dis.*, 64, 102–103.

Davis, J. B., J. S. Brennan, J. Brown, and D. C. Thill, 1994. Quality and grade of canola and canola oil as affected by weed seed contamination, in G. A. Lee, Ed., *Pacific Northwest Canola Conf.: 1994 Proceedings*, University of Idaho, Moscow, 49–52.

Douglas, C. L., Jr., D. J. Wysocki, J. F. Zuzel, R. W. Rickman, and B. L. Klepper, 1990. Agronomic Zones for the Dryland Pacific Northwest, PNW Extension Pub. No. 354, Oregon State University, Corvallis, 8 pp.

Edwards, C. A., 1975. Effects of direct drilling on the soil fauna, *Outlook Agric.*, 8, 243–244.

Funderburk, J. E., D. L. Wright, and I. D. Teare, 1990. Preplant tillage effects on population dynamics of soybean insect pests, *Crop Sci.*, 30, 686–690.

Gregory, W. W., 1974. No-tillage corn insect pests of Kentucky—A five year study, in *Proceeding No-Tillage Research Conference*, University of Kentucky, Lexington, 46–58.

Hagedorn, D. J., Ed., 1984. *Compendium of Pea Diseases*, APS Pres, St. Paul, MN, 57 pp.

Hickman, M., 1995. Personal communications. West Lafayette, IN.

Homan, H. W. and J. P. McCaffrey, 1993. Insect Pests of Spring-Planted Canola, Cooperative Extension CIS 982, College of Agriculture, University of Idaho, Moscow, 4 pp.

House, G. J. and B. R. Stinner, 1983. Arthropods in no-tillage soybean agroecosystems: community composition and ecosystem interaction, *Environ. Manage.*, 7, 23–28.

House, G. J. and B. R. Stinner, 1987. Influence of soil arthropods on nutrient cycling in no-tillage agroecosystems, in G. J. House and B. R. Stinner, Eds., Arthropods in Conservation Tillage Systems, Entomology Society of America Misc. Publ. No. 65, Washington, D.C., 44–52.

Kephart, K. D. and G. A. Murray, 1990. Dryland winter rapeseed production guide, Cooperative Extension Bull. No. 715, College of Agriculture, University of Idaho, Moscow, 23 pp.

Kephart, K. D., M. E. Rice, J. P. McCaffrey, and G. A. Murray, 1990. Spring rapeseed culture in Idaho, Cooperative Extension Bull. No. 681, College of Agriculture, University of Idaho, Moscow, 10 pp.

Koskinen, W. C. and C. G. McWhorter, 1986. Weed control in conservation tillage, *J. Soil Water Conserv.*, 41, 365–370.

Kraft, J. M., D. E. Wilkins, A. G. Ogg, Jr., L. Williams, and G. S. Willett, 1991. Integrated Pest Management Practices for Green Peas in the Blue Mountain Region, Cooperative Extension, Washington State University EB1599, 28 pp.

Lamb, R. J., 1984. Effects of flea beetles, *Phyllotreta* spp. (Chrysomelidae: Coleoptera), on the survival, growth, seed yield and quality of canola, rape and yellow mustard, *Can. Entomol.*, 116, 269–280.

Mathre, D. E., 1982. *Compendium of Barley Diseases*, The American Phytopathological Society, St. Paul, MN, 78 pp.

McCaffrey, J. P., 1990. Insect pests, in K. D. Kephart and G. A. Murray, Eds., Dryland Winter Rapeseed Production Guide, Cooperative Extension Bull. No. 715, University of Idaho, Moscow, 14–18.

Milbrath, L. R., M. J. Weiss, and B. G. Schatz, 1995. Influence of planting date and tillage system of crucifer oilseeds on flea beetle populations (Coleoptera: Chrysomelidae), *Can. Entomol.*, 127, 289–293.

Moore, K. J. and R. J. Cook, 1984. Increased take-all of wheat with direct drilling in the Pacific Northwest, *Phytopathology*, 74, 1044–1049.

Muehlbauer, F. J., W. J. Kaiser, S. L. Clement, and R. J. Summerfield, 1995. Production and breeding of lentil, *Adv. Agron.*, 54, 283–332.

Musick, G. L. and L. E. Beasley, 1978. Effect of the crop residue management system on pest problems in field corn (*Zea mays* L.) production, in *Crop Residue Management Systems*, American Society of Agronomy, Madison, WI, 173–186.

Olkowski, W., H. Olkowski, and S. Daar, 1988. What is IPM? *Common Sense Pest Control*, 4(3), 9–16.

Papendick, R. I., F. L. Young, K. S. Pike, and R. J. Cook, 1995. Northwest: description of the region, in R. I. Papendick and W. C. Moldenhauer, Eds., Crop Residue Management to Reduce Erosion and Improve Soil Quality, U.S. Department of Agriculture-Agriculture Research Service, Conservation Research Rep. 40, Washington, D.C., chap. 3, 4–9.

Peterson, V. F., 1982. A Comparative Survey of Insects in No-Till and Conventional Till Winter Wheat Fields, M. S. thesis, Washington State University, Pullman, 49 pp.

Pike, K., R. Veseth, B. Miller, R. Schirman, L. Smith, and H. Homan, 1993. Hessian fly management in conservation tillage systems for the inland Pacific Northwest, *PNW Conservation Tillage Handbook Series*, chap. 8, No. 15, University of Idaho, Moscow.

Robinson, R. R., A. Retan, and R. Portman, 1974. Insects of Peas, Pacific Northwest Cooperative Extension Pub. PNW 150, Washington State University, Pullman, 19 pp.

Sloderbeck, P. E., 1989a. Brown Wheat Mite, Southwest Kansas Research–Extension Center Circular, January.

Sloderbeck, P. E., 1989b. Winter Grain Mite, Southwest Kansas Research–Extention Center Circular, January.

Smiley, R. W., A. G. Ogg, Jr., and R. J. Cook, 1992. Influence of glyphosate on Rhizoctonia root rot, growth and yield of barley, *Plant Dis.*, 76, 937–942.

Strand, L. L., 1990. Integrated Pest Management for Small Grains, University of Calif. Publ. 3333, Division of Agriculture and Natural Resources, Oakland, 126 pp.

Summerfield, R. J., F. J. Muehlbauer, and R. W. Short, 1982. Description and Culture of Lentils, Production Research Rpt. No. 181, U.S. Department of Agriculture, Washington, D.C.

Thill, D. C., V. L. Cochran, F. L. Young, and A. G. Ogg, Jr., 1986. Weed management in annual cropping limited-tillage systems, in F. L. Elliott, Ed., *STEEP—Conservation Concepts and Accomplishments*, Washington State University, Pullman, 275–287.

Troxclair, N. N. and D. J. Boethel, 1984. Influence of tillage practices and row spacing on soybean insect populations in Louisiana, *J. Econ. Entomol.*, 77, 1571–1579.

Unger, P. W., 1990. Conservation tillage systems, in R. P. Singh, J. F. Parr, and B. A. Stewart, Eds., *Dryland Agriculture: Strategies for Sustainability, Advances in Soil Science*, Vol. 13, Springer-Verlag, New York, 27–68.

Veseth, R., B. Miller, S. Guy, D. Wysocki, T. Murray, R. Smiley, and M. Wiese, 1993. Managing Celphalosporium stripe in conservation tillage systems, PNW Conservation Tillage Handbook Series, Chap. 4, No. 17, University of Idaho, Moscow, 8 pp.

Veseth, R., A. Ogg, D. Thill, D. Ball, D. Wysocki, F. Bailey, T. Gohlke, and H. Riehle, 1994. Managing downy brome under conservation tillage systems in the inland northwest crop-fallow region, PNW Conservation Tillage Handbook Series, Chap. 5, No. 15, University of Idaho, Moscow.

Weller, D. M., R. J. Cook, G. MacNish, E. N. Bassett, R. L. Powelson, and R. R. Petersen, 1986. Rhizoctonia root rot of small grains favored by reduced tillage in the Pacific Northwest, *Plant Dis.*, 70, 70–73.

Wicks, G. A. and B. R. Somerhalder, 1971. Effects of seedbed preparation for corn on distribution of weed seed, *Weed Sci.*, 19, 666–668.

Wiese, M. V., 1987. *Compendium of Wheat Diseases*, 2nd ed., APS Press, St. Paul, MN, 112 pp.

Wiese, M. V., R. J. Cook, D. M. Weller, and T. D. Murray, 1986. Life cycles and incidence of soilborne plant pathogens in conservation tillage systems, in L. F. Elliott, Ed., *STEEP—Conservation Concepts and Accomplishments*, Washington State University, Pullman, 299–313.

Wiese, M. V., W. J. Kaiser, L. J. Smith, and F. J. Muehlbauer, 1995. *Ascochyta* blight of chickpea. Univ. of Idaho Coop. Extension System, Agric. Experiment Station, Current Information Series 886. 4 pp.

Williams, L. III, 1993. Studies of the Pea Leaf Weevil (*Sitona lineatus* L.) and *Limonius* spp. Wireworms in a Pea, *Pisum sativum* L./Wheat, *Triticum aestivum* L., Integrated Crop Management System, Ph.D. dissertation, University of Idaho, Moscow, 131 pp.

Williams, J. L. and G. A. Wicks, 1978. Weed control problems associated with crop residue systems, in W. K. Oschwald, Ed., Crop Residue Management Systems, ASA Spec. Publ. 32, American Society of Agronomy, Crop Science Society of American, and Soil Science Society of America, Madison, WI, 165–172.

Young, F., R. Veseth, D. Thill, W. Schillinger, and D. Ball, 1995. Managing Russian Thistle under Conservation Tillage in Crop-Fallow Rotations, Pacific Northwest Extension Pub. PNW-492, University of Idaho, Moscow, 12 pp.

Young, F. L., A. G. Ogg, Jr., D. C. Thill, D. L. Young, and R. I. Papendick, 1996. Weed management for crop production in the Northwest wheat (*Triticum aestivum*) region, *Weed Sci.*, 44, 429–436.

Zehnder, G. W. and J. J. Linduska, 1987. Influence of conservation tillage practices on populations of Colorado potato beetle (Coleoptera: Chrysomeliade) in rotated and non-rotated tomato fields, *Environ. Entomol.*, 16, 135–139.

7 Developments in Equipment for Conservation Farming

Charles L. Peterson

CONTENTS

The STEEP (Solutions to Environmental and Economic Problems) program was initiated in 1975 to address soil erosion and related crop production problems in the Pacific Northwest. Existing tillage methods were major contributors to the erosion problem and consequently were one thrust of the STEEP program. It is generally acknowledged that the majority of erosion in the Palouse region occurs on land

0-8493-1185-3/99/$0.00+$.50
© 1999 by CRC Press LLC

planted to winter wheat. The relatively high season precipitation and very wet soils combined with the common practice of preparing a fine seedbed on the long, steep slopes of the area lead to excessive winter season runoff and soil loss. It has shown that erosion can be reduced as much as 75% over conventional tillage methods when minimum-tillage planting concepts are used to plant fall wheat onto land which was previously cropped to peas or lentils.

Water conservation is also improved using minimum-tillage seeding. Data collected by the University of Idaho, Department of Biological and Agricultural Engineering indicate that 1.8 times more precipitation is lost as surface runoff from conventional tillage seeding as from minimum-tillage seeding. Total nitrogen and phosphorus losses from reduced-tillage seeding were reduced significantly when compared with conventional tillage seeding methods.

In addition to reducing soil, water, and nutrient losses, minimum-tillage seeding also saves field time and fuel. The field time requirement has been shown to be 18 to 47% less and diesel fuel requirements 50 to 70% less when the one-pass method of seeding is used compared with two conventional seeding systems. Another advantage of one-pass seeding is that the tractor pulling the equipment operates on previously undisturbed soil, which will allow the machine to be operated sooner after a rainstorm and earlier in the spring than would be possible with similar-sized equipment operating on tilled soil. Thus, more days are available for field work, which allows the grower to expand to more acres or to be a more efficient manager of present acres.

Morrison et al. (1988) reported that in 1986 there were 44 kinds of planters and 121 kinds of drills and air seeders available in the U.S. for conservation seeding. In addition, many add-on components were available from specialty companies from which total machines could be constructed using components from many sources. They suggest that planting machines can be characterized by their components that actively engage the soil. These components are used to perform the following machine functions:

1. Soil and residue cutting
2. Row preparation
3. Soil opening for seed placement
4. Firming uncovered seed
5. Seed covering
6. Seed furrow closure and firming
7. Depth control

Table 7.1 is a list of planting machine components that might be used to perform each of the seven functions. Figure 7.1 is an example of soil-engaging component sequence on a conservation tillage seeding machine with components selected from Table 7.1. Through the research efforts of several scientists and small-farm-machine manufacturers, major gains have been made in reducing erosion through developments in minimum-tillage and no-tillage systems. Some of these developments included: new and improved furrow openers; methods to apply fertilizer to achieve highest use

TABLE 7.1
Planting Machine Components for Each of Seven Machine Functions

1. Soil and Residue Cutting
   a. Smooth coulter
   b. Notched coulter
   c. Coulter with depth bands
   d. Bubble coulter
   e. Rippled coulter
   f. Narrow fluted coulter
   g. Wide fluted coulter
   h. Straw straightener
   i. Powered blade or coulter
   j. Strip rotary tiller
   k. Dual secondary residue disks
   l. Not used
   m. Provided by the front unit
2. Row Preparation
   a. Sweep row cleaner
   b. Vertical disks row cleaner
   c. Horizontal-disk row cleaner
   d. Wide fluted coulter loosener
   e. Hoe opener
   f. Single-disk opener
   g. Coulter opener
   h. Chisel opener
   i. Wide-sweep opener
   j. Triple-disk opener
   k. Powered blade or coulter
   l. Provided by the rear unit
3. Soil Opening for Seed Placement
   a. Double-disk opener
   b. Staggered double-disk opener
   c. Runner opener
   d. Stub-runner opener
   e. Hoe opener
   f. Single-disk opener
   g. Coulter opener
   h. Chisel opener
   i. Wide-sweep opener
   j. Triple-disk opener
   k. Powered blade or coulter

4. Seed Firming
   a. Semipneumatic wheel
   b. Solid wheel
   c. Not used
   d. Provided by the rear unit
5. Seed Covering
   a. Single covering disk
   b. Double covering disks
   c. Covering paddles
   d. Covering knives
   e. Covering chains
   f. Spring covering tines
   g. Not used
   h. Provided by the rear unit
6. Seed Slot Closure
   a. Wide semipneumatic press wheel
   b. Wide steel press wheel
   c. Single-rib semipneumatic press wheel
   d. Double-rip semipneumatic press wheel
   e. Narrow semipneumatic press wheels; "V", rounded
   f. Dual-angled semipneumatic press wheels
   g. Dual-angled cast or steel press wheels
   h. Narrow steel press wheel; "V", rounded
   i. Split narrow press wheels
   j. Dual wide press wheels
   k. Not used
   l. Provided by the rear unit
7. Depth Control
   a. Rear press wheel
   b. Side gauge wheel
   c. Skid plate
   d. Front wheels and rear press wheels tandemed
   e. Frame lifting gauge wheels
   f. Depth bands on front
   g. Depth bands on disk opener
   h. Downpressure only
   i. Provided by the rear unit

Source: Morrison, J. E., Jr. et al., *Trans. ASAE*, 32(2), 397–401, 1989. With permission.

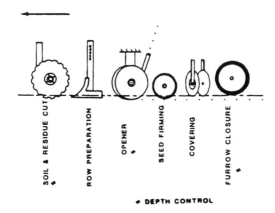

FIGURE 7.1 Example of a soil-engaging component sequence on a conservation tillage seeding machine. The depth control component may be located at any one or more of the locations indicated. (From Morrison, J. E., Jr. et al., *Trans. ASAE*, 32(2), 397–401, 1989. With permission.)

efficiency; construction and testing of complete machine systems; and consideration of totally new farming systems.

7.1 NEW AND IMPROVED OPENERS

Development of new and improved openers focuses on trash clearance, seed placement, fertilizer placement, and soil firming. Placement of starter fertilizer near the seed row and/or below the seed has received much attention by researchers developing openers for conservation tillage (for example, Payton et al., 1985). Examples of the various opener types are shown in Figure 7.2 taken from Wilkens et al. (1983). A brief description of openers developed by STEEP-sponsored researchers is given in this section.

7.1.1 SHANK-TYPE OPENERS

Wilkins (1988) and others at the Columbia Research Conservation Center experimented with deep-furrow grain drill openers, modified for banding fertilizer below the seed. These openers, with some variations, are available commercially in the Pacific Northwest. The original models disturbed too much soil and led to research with a narrow opener that created less soil disturbance and reduced draft. The narrower patented design, shown in Figure 7.3, was designed to place liquid fertilizer 51 mm (2 in.) below the seed. The knife point is mounted on a John Deere HZ deep-furrow opener shank. The knife cuts a slot approximately 6.3 mm (0.25 in.) wide and 90 mm (3.5 in.) deep. Above the knife point, the opener is 32 mm (1.25 in.) wide for moving the dryer surface soil away from the seed row.

One type of opener tested by Wilkins and Haasch (1990) minimizes the soil disturbance by making two furrows. The first furrow in which fertilizer is placed is

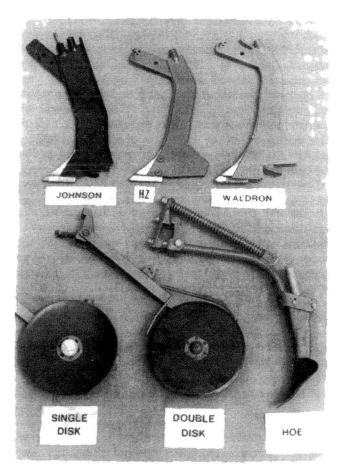

FIGURE 7.2 Examples of opener types. (From Wilkins, D. E. et al., Spec. Rep. 776, Agricultural Experiment Station, Oregon State University, Pendleton, 1983. With permission.)

very narrow; a second and wider furrow is made directly over the fertilizer furrow. Experimental results indicated this type of opener had promise for seeding small grain in the Columbia Plateau. A commercial opener made by S and M Manufacturing Co. uses the two furrow concept, Figure 7.4.

Hyde and Simpson at Washington State University (Veseth, 1985) evaluated a parabolic knife opener (Figure 7.5). This opener is 9.5 mm (3/8 in.) wide with a 19 mm (3/4 in.) wear point on the bottom. The opener is reportedly capable of placing anhydrous ammonia, liquid fertilizer, dry fertilizer, and seed. The parabolic knife opener reduced the amount of soil disturbance and power requirements under no-till seeding compared with the wide John Deere HZ openers. The parabolic shape of the opener helps to clear residue better than the straighter HZ opener, enabling the drill to seed no-till through more residue without plugging.

Peterson et al. (1983) at the University of Idaho developed a modified chisel shank opener for placing liquid fertilizer below the seed. Chisel points, 51 mm (2 in.)

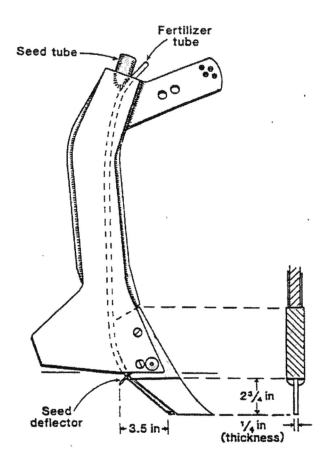

FIGURE 7.3   Modified deep-furrow drill openers. (From Wilkins, D. E., U.S. Patent 4,765,263, 1988. With permission.)

wide, opened furrows, which were then seeded with double-disk openers. The wide shank disturbed too much soil, often resulting in nonuniform seeding depth. Seed rows on the ridges tended to be planted too deeply and seed rows in the deeper furrows often had rodent damage. They then designed a twisted shank. The 51-mm (2 in.)-wide shank was twisted 90°, as shown in Figure 7.6, which resulted in a 19-mm (³/4-in.)-wide furrow opener. Commercial 19-mm (³/4-in.), hardened knife points were added to improve wear and soil penetration. The narrow shank opener significantly improved seed germination and survival, while reducing the amount of soil disturbance and power requirement.

7.1.2   DISK OPENERS

Openers from shanks can pass less residue and stubble than disk-type openers. However, disk openers have more difficulty penetrating the firmer soil encountered in no-till seeding. Disk openers considered for no-till have included single-disk,

FIGURE 7.4 S and M Manufacturing Company grain drill opener point. (From Wilkins, D. E. and Haasch, D. A., Columbia Basin Agricultural Research Spec. Rep. 860, Oregon State University, Pendleton, OR, 1990.)

double-disk, and special heavy-duty disk openers designed by various manufacturers. Heavy-duty disk openers generally have larger disks, heavier bearings, and are attached to a heavy-duty frame to provide the downward thrust forces required to force the disks into firm soil. Some manufacturers stagger two disks to aid penetration. Disk openers range in size from 152 mm (6 in.) in diameter to 610 mm (24 in.) or more in diameter. Downward weights required vary from 45 kg (100 lb) per opener to over 225 kg (500 lb) per opener depending on the manufacturer and soil conditions. Smaller openers generally require less downward force. Some machines make the downward thrust force variable by various mechanical or hydraulic mechanisms in order to fit the particular soil conditions. Examples of disk openers will be shown with the various machines.

Disk drills use a variety of press wheel styles for soil firming and depth gauging. Disk openers are either a single-disk or a double-disk type and may or may not have a coulter ahead of the openers to start the soil penetration process and to cut residue. Coulters are of various designs including smooth, notched, rippled, and fluted.

Fertilizer placement with disk openers may be with the seed; slightly deeper and ahead of the seed with a soil layer for separation; or in separate furrows with disks or shanks used only for fertilizer placement.

7.2 MINIMUM-TILL AND NO-TILL PLANTERS

Openers are the key to successful planting in minimum-tillage or no-till systems. Even so, to be useful, openers must be successfully integrated into a complete machine. A single opener may have every desirable feature but fail to work when multiple openers are attached together on a tool bar or frame. A successful seeding mechanism consists of multiple openers with the necessary seed boxes, fertilizer hoppers or tanks, metering systems, and raising and lowering mechanisms all

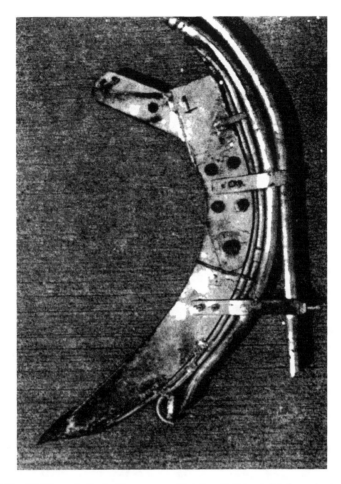

FIGURE 7.5   WSU parabolic knife opener with tubes (left to right) for anhydrous ammonia, liquid and dry fertilizer, and seed.

attached to a frame-and-hitch system. The machines must be capable of withstanding the stresses of planting in undisturbed soil, light enough to avoid excessive compaction, nimble enough for transport from field to field, and low enough in cost to be within the reach of a significant number of farmers.

Drills can be classified into several categories:

- By tillage type:
  Minimum tillage
  No-till
- By opener type:
  Hoe
  Chisel
  Disk

FIGURE 7.6   Twisted chisel shank opener used by Peterson et al. (1983).

- By fertilizer type:
  Dry
  Liquid
  Anhydrous
- By implementation:
  Research
  Commercial

The following section is a description of some of the drills tested as part of STEEP research. Not every drill manufacturer or research drill is shown; however, several types of drills and styles are shown to illustrate the major categories of drills tested.

7.2.1   THE YIELDER

Originally called the "Pioneer," this large, no-till drill deserves special attention because of its originality in design, massive features, and ability to plant in high-residue conditions. The Yielder was a true no-till system (Swanson, 1983a, b). It had both fertilizing and seeding capability and planted into nondisturbed residue. One model of Yielder at work on a Palouse hillside is shown in Figure 7.7.

Yielder Floating Tool Bar, Patent 4,333,534 (Figure 7.8)—Allowed the ground-engaging tools to follow the terrain independently of the other tool bars. Depth control was maintained by the front depth wheel and the rear packer wheel. A hydraulic cylinder applied load to the tool bar. Pressure applied to the tool bar could

FIGURE 7.7   The Yielder drill manufactured in Spokane, Washington.

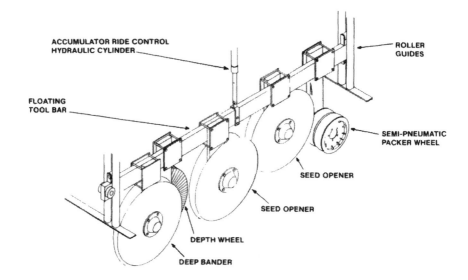

FIGURE 7.8   The floating tool bar used on the Yielder drill.

be varied by the operator. The pressure, once set, remained constant since the oil is transferred to a hydraulic accumulator.

Yielder Seed Opener, Patent 4,044,697 (Figure 7.9)—The "Swanson" offset leading double-disk released seed into the opening convergence of the disk. As the disk turned, the seed was thrust downward into the vee. Another unique feature of the opener was the different relative motions between the two disks, which helped for more effective straw cutting. An operator-controlled hydraulic cylinder provided a constant vertical force to the opener to place seed more evenly.

Yielder Deep-Bander (Figure 7.10)—The deep-bander applied up to three different fertilizer materials simultaneously with very little soil disturbance up to 190 mm (7.5 in.) in depth. Yielder claimed that the technique fed the crop and starved the weeds and also aided root colonization deep in the soil profile.

Paired Row—Paired row seeding was a Yielder concept (also referred to as *twin row* or *skip row*). Paired row is described as a close alignment of two seed rows

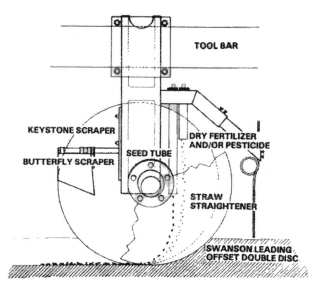

FIGURE 7.9   The Yielder seed opener.

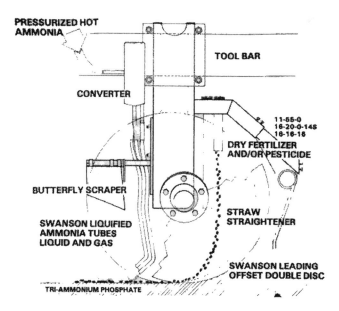

FIGURE 7.10   The Yielder deep-bander showing how three different fertilizer materials could be applied simultaneously.

with a fertilizer deep-band located exactly between and below the close center of the two seed rows. A wide, nonfertile band resulted between the two sets of paired rows. Yielder claimed the paired row to be a most significant development for the following reasons:

FIGURE 7.11 The USDA III drill used for state-of-the-art research on no-till in the Palouse.

1. Deep-banding of fertilizer;
2. More standing stubble;
3. Quicker emergence as a result of midrow packing and more accurate depth control;
4. More residue-handling capability.

7.3   OTHER NO-TILL DRILLS

7.3.1   USDA-ARS, PULLMAN, WASHINGTON

This small, 2.4-m (8-ft) test plot drill, Figure 7.11, was an applied research machine that scientists used to determine the complex interactions of conservation farming. The drill was built by Yielder Mfg. to acquire research data. This drill and the USDA versions I and II, which preceded it, were major contributors to knowledge about no-till farming on dryland wheat production and were instrumental in the development of the Yielder drill. Much of the research data acquired from the use of this drill applied to Yielder drill owners. This state-of-the-art machine had five dry compartments, five wet compartments, and a pressure vessel. It could apply all known fertilizer materials at the time of seeding in various row spacings. Several packer wheel combinations were available.

7.3.2   HAYBUSTER

Haybuster drills, Figure 7.12, were lighter and less expensive than the Yielder drills. They were manufactured in both disk, Figure 7.13, and hoe configurations, Figure 7.14 (Haybuster, 1984).

The Haybuster Hoe Drill—Designed to deep-band fertilizer, it could also place the fertilizer with the seed. Haybuster claimed that research had shown that deep-banding below the seed helped plants get off to a fast start by putting nutrients in the root zone and allowing the roots to use these nutrients as they grow down.

FIGURE 7.12 The Haybuster conservation tillage drill manufactured by Haybuster Mfg., Inc. This drill was available with a variety of tillage and seeding tools, including both disk and hoe configurations.

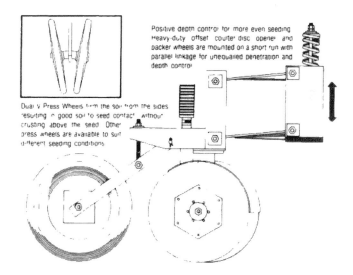

FIGURE 7.13 The Haybuster offset coulter-disk opener and packer wheels were mounted on a short run with parallel linkage for penetration and depth control. Dual V press wheels firmed the soil, resulting in good soil-to-seed contact; other press wheels were also available.

According to Haybuster, deep-banding increased yields by 335 to 1000 kg/ha (5 to 15 bu/acre). The drill had the option of drilling only seed or just fertilizer.

Spring-loaded shanks with replaceable tips and spouts placed the seed in two rows, 51 mm (2 in.) apart and deposited fertilizer 38 mm (1.5 in.) below the seed. These drills were available in 203-mm (8-in), 241-mm (9½-in.), 305-mm (12-in.), and 356-mm (14-in.) row spacings. If desired, fertilizer could be drilled with the seed. The drill could also be used for fall fertilizer banding without seeding or to drill seed only.

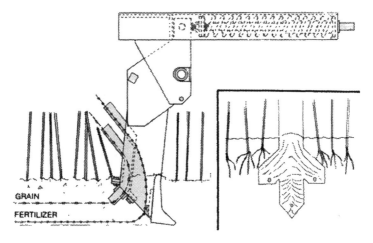

FIGURE 7.14 Haybuster hoe-type openers can be used to place fertilizer with or below the seed or either separately.

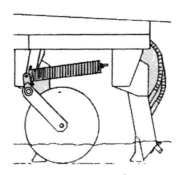

FIGURE 7.15 Haybuster spring-loaded openers running ahead of the hoe-type openers.

Optional spring-loaded 406-mm (16-in.) coulters run ahead of each opener to permit use of the drill in minimum- or no-till conditions, Figure 7.15.

7.4 SHANK AND SEED

7.4.1 AIR SEEDERS

Morrisson et al. (1988) define air seeders as machines that consist of two machine units pulled in tandem. The first machine has a central hopper with a meter that discharges seed into an air distribution and delivery system. The second machine is a common tillage implement such as a chisel plow, field cultivator, or stubble mulch plow. A metering mechanism and seed tubes on the first machine deliver the seed to the soil openers on the second machine. Various soil openers have been used, including chisel points, chisel sweeps, large V-blades, and some of the special openers discussed earlier. Many of these drills also provide press wheels or other covering and firming mechanisms behind the furrow openers.

Prasco Super Seeder—There were two Prasco Super Seeder models (Prasco, 1980). The Super Seeder 125 had a 4400-l (125-bu) grain hopper with a single large filling port, to load faster and last longer in the field. It was equipped with a Dempster ground drive fertilizer pump. Super Seeder 65 had a 2290-l (65-bu) grain tank, plus a 1136-l (300 gal) liquid fertilizer tank and was also equipped with a ground drive fertilizer pump. The Prasco Super Seeder system is shown in Figure 7.16. The Prasco used a single-seed-metering system to place the seed into the airstream and a distribution system to subdivide the seed stream into each individual seed tube. Seed was deposited in the soil behind cultivator sweeps.

7.4.2 Air Drills

Morrison et al. (1988) defined air drills as machines with central seed hoppers, integral seed metering, and air distribution and delivery systems, similar to those used for air seeders. The difference between air drills and air seeders is that air drills have hoe or double-disk furrow openers as used on other types of drills. According to Morrison et al. (1988), individual row unit suspensions and depth-controlling press wheels follow soil surface contours and control seed depth better than air seeders.

Chisel-Planter—The Chisel-Planter, Figure 7.17, minimum-tillage system, was developed by combining three successful practices common to the Palouse; use of the chisel plow for runoff control, fall-applied fertilizer incorporated into the seed-bed, and use of the fluted-feed, end wheel drill. The Chisel-Planter is properly classified as a till–plant machine since chisels work the soil directly ahead of seed placement. The chisels (Figure 7.2) are placed 610 mm (24 in.) apart, staggered in two ranks to give a 305 mm (12 in.) row spacing. Liquid fertilizer is metered into the furrow at the bottom of each chisel point. The chisels are designed to run at the 100 to 130 mm (4 to 5 in.) depth. The original Chisel-Planter used packer wheels of 152-mm (6-in.) diameter attached directly to the chisel shanks. They provided 25 to 50 mm (1 to 2 in.) of compacted soil cover between the fertilizer and seed and also broke up large clods, thus improving the seedbed directly in the row. The packer wheels were removed when narrow shanks were added and resulted in much-improved trash clearance. The seed, metered with a standard fluted feed assembly, is carried pneumatically to the double-disk openers, where it is placed into the furrow opened by the chisel. Small rubber packer wheels follow the seed openers on the Chisel-Planter providing a final firming of the seedbed to enhance germination.

The Chisel-Planter method of seeding leaves ridges in the field that might be categorized as "miniterraces." It is this miniterracing effect coupled with surface residue retention that provides the excellent erosion control.

The first model of the Chisel-Planter built in 1974 was a single shank mounted on a three-point hitch tool bar for testing the concept of planting in the furrow following a chisel opener. The next model was an unwieldy machine, 5.2 m (17 ft) wide, made from a chisel plow and end wheel drill with three ranks of modular chisel/opener combinations. This machine was quickly found to have difficulty

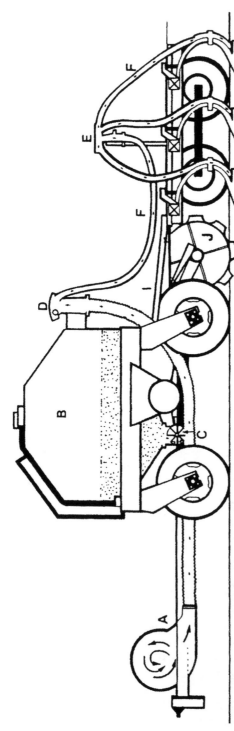

A. Blower - Driven by a hydraulic motor to be run from a tractor hydraulic system or a separate hydraulic system.

B. Grain Tank - Hopper bottom. One fill hole. Clean-out door.

C. Metering Device - Meters all types of grain accurately into the air stream.

D. Primary Manifold - Distributes grain accurately to the secondary manifolds.

E. Secondary Manifold - Distributes grain accurately to each shank.

F. Flexible Tubing - Carries grain. This allows wings to be folded with no alterations to seeding system.

G. Boot - Plastic boot is located behind shank to protect it from rocks.

H. Liquid Fertilizer Holder - Holds liquid fertilizer tubes in place behind seed tube.

I. Rolling Hitch - Allows chisel plow to follow contour of ground for accurate depth control.

J. Ground Drive Wheel - Drives seed metering device and liquid fertilizer pump with no clutches. Wheel is lifted off ground when chisel plow is raised.

FIGURE 7.16 A schematic and flow diagram of the Prasco Super Seeder.

FIGURE 7.17   University of Idaho Chisel Planter. (From Peterson, C. L. et al., *Trans. ASAE*, 26(2), 378–383 and 388, 1983. With permission.)

clearing trash. Also, the most rearward chisels threw soil, deeply burying the seed planted by the more forward modular units, causing poor emergence. While neither of these tests were completely successful, they did provide the background for development of the present Chisel-Planter. The successful parts of the early machine, i.e., delivering the seed with an air injection system and placing liquid fertilizer at the tip of the chisel point below the seed, were incorporated with a more acceptable chisel and opener arrangement for tilling, fertilizing, and planting in one pass.

One of the benefits of the Chisel-Planter is that residue from the previous crop is cleared from the furrow, ahead of the opener. The adverse effects of decomposing crop residues on plant growth are widely recognized and have been discussed in detail in STEEP reports. This toxic effect of crop residue and the deleterious microbial/residue interaction was of concern during the development of the original minimum-tillage concepts and contributed to the selection of equipment.

The last major change to the Chisel-Planter occurred in 1984 when the 51 mm (2 in.) chisel shanks, Figure 7.18 (top), were replaced with narrow shanks, Figure 7.18 (bottom). These shanks have hardened narrow shovels on the tips and fertilizer tubes behind. Total draft reduction through use of the narrow shanks was estimated at 25%. The narrow shank provided a more uniform seeding depth and even miniterraces or ridges across all rows. Erosion and yield benefits were difficult to estimate, but no dramatic change in performance as measured by yield, emergence, or soil erosion could be reported. The narrow shanks did have the effect of less soil disturbance and more residue left on the surface.

Since the Chisel-Planter was developed in 1975, it was used in numerous trials, drill comparisons, seeding tests, etc., throughout Idaho and eastern Washington. In

FIGURE 7.18   Original 51-mm (2-in.)-wide vibra-chisels with packer wheel additions (top) and narrow, twisted shanks with hardened point used later on the University of Idaho Chisel-Planter. (From Peterson, C. L. et al., *Trans. ASAE*, 26(2), 378–383 and 388, 1983. With permission.)

recent years, it has become a standard by which various drill comparisons can be evaluated.

7.4.3   CHISEL-PLUS-DRILL

The need for a low-cost reduced-tillage seeding system similar to the Chisel-Planter but which could be easily duplicated by growers for use on their own farms was the basis upon which the chisel-plus-drill was developed, Figure 7.19. The chisel-

FIGURE 7.19 The chisel-plus-drill developed for experimentation with low-cost minimum tillage.

plus-drill was two implements in tandem. The chisel plow was equipped with an engine-driven fertilizer pump, tank, and injection system and pulls an end wheel grain drill with double-disk openers. Liquid fertilizer (usually a combination of solution 32, 10-34-0, and thio-sol) was injected into the soil 100 to 150 mm (4 to 6 in.) deep with the chisel plow shanks. Spacing of the shanks was 305 mm (12 in.). The chisel-plus-drill placed the seed in 152 mm (6 in.) rows using the double-disk openers of a standard, unmodified end wheel drill.

The chisel-plus-drill was first used in 1982 and has been used in north Idaho under favorable conditions. The chisel-plus-drill leaves the soil surface with 65 to 75% of the residue and moderate-sized clods. Water infiltration is increased and erosion decreased compared with conventional tillage. Other options used with the chisel-plus-drill are 356-mm (14-in.) sweeps and a rod-weeder attachment following the chisel plow.

7.4.4 ADAPTATIONS OF THE CHISEL-PLANTER CONCEPT

A number of adaptations of the Chisel-Planter concept have been built both by commercial firms and local farmers. No information is available on the total number of units sold or total acres seeded. Most of these incorporate improvements and modifications originated with the builder. Figure 7.20 shows a typical example of one such machine in operation.

7.5 DRILL COMPARISONS WITH MINIMUM TILLAGE

A summary of 40 dryland drill comparisons in which no-till drills, the Chisel-Planter, and chisel-plus-drill have participated in the Palouse is shown in Tables 7.2 and 7.3 (Peterson et al., 1988.) Almost all of the tests were on land for which the previous crop was peas or lentils. Sites ranged from 2 to 6 ha (5 to 15 acre). Comparisons were made by selecting adjoining sites of more or less similar topography. Yields were determined by hand harvesting eight to ten subplots within the larger main plots. Erosion was determined by collecting rill data. A summary of the data is listed below.

FIGURE 7.20   One of several farmer-built copies of the Chisel-Planter concept.

TABLE 7.2
Yields from Drill Comparisons 1978 to 1986 (data given as a percent of conventional)

	No. of sites	% of Conventional	Range of Variability	
			% of Conventional (max)	% of Conventional (min)
Chisel-Planter	38	95.2	159.8	38.9
Chisel-plus-drill	7	93.2	132.3	72.8
No-till[a]	10	94.3	154.3	76.1
Conventional	40	100.0	100.0	100.0

[a] Various no-till drills were included which were in use on the particular farm used as a test site.

TABLE 7.3
Soil Loss from Drill Comparison Sites 1978 to 1986 (data given as percent of conventional)

	No. of sites	% of Conventional	Range of Variability	
			% of Conventional (max)	% of Conventional (min)
Chisel-Planter	38	36.6	102.0	0.0
Chisel-plus-drill	7	25.7	37.5	0.0
No-till[a]	10	55.7	900.0	0.1
Conventional	40	100.0	100.0	100.0

[a] Various no-till drills were included which were in use on the particular farm used as a test site.

The data show that no-till and minimum tillage reduced yields by 5 to 7% and erosion by 45 to 65%.

7.6 FERTILIZER PLACEMENT

To obtain the most efficient utilization of commercial fertilizer, several factors need to be considered, including

1. Type of fertilizer to use
2. Amount to apply and projected yield
3. Time of application
4. Soil temperature, moisture, and pH
5. Soil type
6. Soil test results
7. Cropping and fertility history
8. Strategic location of the fertilizer within the soil profile.

These factors are interrelated in such a way that optimizing fertilizer use will include consideration of each factor. Fertilizer placement should be considered as only one facet of a complex problem. Proper choice of placement must be made in the context of the total situation.

There are many possible placements for fertilizer, all of which have their advocates for a particular cropping system; these include

Broadcast—A uniform application to the soil surface;

Strip Broadcast—A narrow band or strip of fertilizer placed on the soil surface;

Strip Broadcast with Plowdown—Used to mix a concentrated source of nutrient within a small percentage (5 to 20%) of the soil profile;

Starter or Pop-up—A small amount of fertilizer placed in the seed row;

Banded—Fertilizer placed below the soil surface in a concentrated strip;

In-the-Row—Placing the fertilizer in the row with the seed;

Deep-Banded—Placing the fertilizer band well below the seed depth, 10 cm or more;

Side-Banded—Placing the fertilizer bands to the side of the seed row, below the soil surface at, above, or below seed depth;

Side-Dressed—Placing the fertilizer in bands below the soil surface after the crop is growing;

Paired Row—Placing a single band of fertilizer below the surface, equidistant between two seed rows;

Nested—Placing fertilizer intermittently down the seed row in a concentrated clump or "nest" (90% of the fertilizer in 25% of the internest spacing); nests may be 15 to 40 cm apart along the row;

Super Granules—Urea granules of 1, 2, or 3 g;

Large, Multiyear Applications—A large amount of fertilizer is broadcast on the surface once every few years.

The various placements have all been used successfully for specific purposes. For this reason, the literature appears to contain inconsistencies. One needs to analyze the total context in which the study was made before reaching a conclusion about its applicability to a particular situation.

The following specific conclusions were noted as a result of fertilizer placement studies (Peterson et al., 1988):

1. A significant difference in yield due to placement was not detected. However, fertilizer with the seed tended to be associated with lower overall yields.
2. Sulfur placed with the seed decreased emergence and yield at two locations when soil moisture was relatively low.
3. With phosphate, the first increment (22.4 kg/ha; 20 lb/acre) produced virtually all of the yield increase observed.
4. No increase in yield due to rate of sulfur applied was noted.
5. Spring-applied nitrogen was more efficient than fall-applied nitrogen; i.e., the yield increase per unit of applied N was higher.

7.7   OTHER CONSERVATION TILLAGE EQUIPMENT

7.7.1   PARAPLOW

An innovative tillage implement termed Paraplow (a registered trademark of the Howard Rotavator Co.) was developed in recent years in England by the cooperative effort of university and private industry (Pigeon, 1983).

The result was the development of a "slant-leg" chisel, which would raise and shatter the soil to a significant working depth, while doing minimal disturbance to the soil surface and surface residue. Although not a plow, it did create a "plowlike" disturbance to the near surface soil, thus was dubbed a "paraplow." Using multiple shanks on a plow-type frame allows full field working with a draft slightly greater than the traditional moldboard plow but requiring virtually no subsequent tillage before direct seeding.

Saxton et al. (1986) described the Paraplow tillage action as somewhat unique when compared with other traditional tillage implements. The slant-leg chisel runs at a working depth of 356 to 406 mm (14 to 16 in.) with an adjustable shatter plate at the bottom to create a variable amount of lifting and shattering. The soil is lifted over the angled chisel and fractured, thus creating multiple cracks and additional porosity without significant soil surface disturbance. This soil loosening has the effect of creating macroporosity connected to the soil surface, which can increase water infiltration even under very wet or frozen soils. This soil loosening will allow much easier subsequent tillage and plant root penetration. Wilkins et al. (1986) found that the Paraplow tillage treatment shattered the deep tillage pan and reduced the cone index by more than 30%.

The preliminary results of research on the characteristics and performance of the Paraplow were obtained by researchers at the USDA-ARS, Pullman, Washington and their cooperators. Through the use of an air infiltrometer, they were able to show that

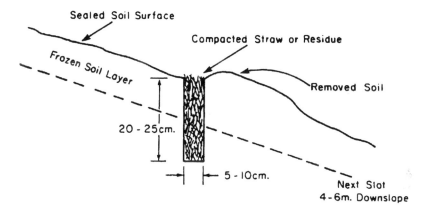

FIGURE 7.21 Schematic of slot-mulch treatment on cereal crop residue. (From Hyde, G. M. et al., *Trans. ASAE*, 29(1), 20–25, 1986. With permission.)

this implement would create a stable, surface-connected macroporosity effective in maintaining good infiltration characteristics throughout the winter period, even though several deep freezes occurred. Runoff measurements from Paraplow-tilled plots confirmed that very little runoff and erosion occurred even under rather severe weather conditions. Several sets of penetration measurements at different plot locations following different tillage sequences and with various crops have been conducted.

Saxton et al. (1986) report that, without exception, the Paraplow-tilled plots have shown much-improved soil "looseness," as would be expected with the obvious soil shattering, particularly when the tillage is done on dry soil conditions. In most cases, the test plots had been directly seeded following Paraplow tillage either immediately in the fall or as spring-seeded no-till. With few exceptions, no major problems occurred and good stands resulted. In some cases, the drill had to be readjusted to account for the much softer soil surface condition than in the non-Paraplow-tilled condition, especially where a heavy, no-till-type drill was being used. A standard double-disk drill was successfully used in several cases where light to moderate residue was present.

7.7.2 SLOT-MULCH

Hyde et al. (1986) reported on a system for creating what they call "slot-mulch." The system, similar to that tried by Moden (1965), creates 50 to 100 mm (2 to 4 in.) wide by 200 to 250 mm (8 to 16 in.) deep soil slots and then compacts them full of crop residue, Figure 7.21. The slots are cut on the contour as much as 4 to 6 m (13 to 20 ft) apart. Hyde et al. (1986) report that the crop residue packing prevents freezing near the bottom of the slot, water flow along the slot, and surface sealing over the slot. The slots channel runoff water below the crusted and frozen soil layers that often exist during midwinter, when much of the runoff and erosion occur in the region. Field trials showed that soil losses were 0.02 to 0.09 t/ha (0.04 ton/acre) for slot mulch compared to 1.12 t/ha (0.5 ton/acre) for the no-tillage plot and 26.0 t/ha (11.6 ton/acre) on an adjacent summer fallow plot.

7.8 PRECISION FARMING

Precision farming, spatially variable agriculture, prescription farming are all names that have been used to identify a management concept whereby large fields are divided into small cells and, for management, each cell is treated as an individual. The technique is made possible through electronic controls; small, high-speed, inexpensive computing systems; and the ability to locate the machine in the field, on-the-go, through use of the Global Positioning System (GPS). For example, a 32.4-ha (80-acre) field divided into 40 m (130 ft) by 40 m (130 ft) cells requires 202 individual management units. Divisions of this size become too complex for the operator to handle manually. However, the addition of automatic controls functioning via a computer database to control the set point, requires no more effort than applying the input at a uniform rate.

7.8.1 SPATIALLY VARIABLE FERTILIZER APPLICATION

STEEP research funded a project to develop a precision fertilizer management system for use in the Palouse (Peterson et al., 1993). The following are from a summary of that project.

Anticipated advantages of the accurate placement of fertilizer are

1. Fewer productive areas receive less fertilizer, thus less is wasted and less is available for infiltration into the groundwater or to be transported with surface runoff.
2. Problem areas can be fertilized according to the need. For example, in the Palouse, it is known that more fertilizer is lost by movement with groundwater and surface water from the steep slopes. These areas can be underfertilized during the normal fall application and the difference made up with a spring application. The result is a more efficient use of the fertilizer applied and less potential pollution.
3. More productive areas receive more fertilizer. Yields in these areas are enhanced raising the overall average yield of the field. Spatially variable fertilizer techniques do not necessarily reduce the total fertilizer applied, but they do distribute the chemical more advantageously.

The elements of a spatially variable fertilizer application system include

1. A technique for determining the correct application rate for each cell and for finding the best cell size;
2. Creation of a database based on location parameters consisting of the correct rate for each cell;
3. A locating system to find the position of the applicator in the field and thus the current cell;
4. A sensor that will send a signal proportional to implement ground speed to the mobile computer;
5. A mobile computer system located in the applicator that can access the field position system, the application rate database, and a speed sensor;

use this information to compute the correct bias for the applicator control system; and put this control signal on the computer bus;

6. A chemical applicator with a rate control system that can receive a signal from the mobile computer system and use that signal for adjusting the application rate set point;

7. Provisions for letting the farm manager override the system and to have input on the fertilizer rate found for each cell. This assures that the manager is still in control of the production system.

Of the seven elements listed, items 2 through 7 are relatively straightforward using current technology. Item 1, determining the rate of fertilizer required for each cell, is the most difficult, and considerable research remains in developing the most efficient procedures.

To propose using satellite signals to control a farm machine, on-the-go, in the field may seem hypothetical or "blue sky." Each of the steps described has been carried out on an actual field-scale experiment. The fertilizer was applied as reported and the equipment required is available at reasonable cost. Determination of the actual on-site benefits is a more difficult problem and will take considerable additional research effort to quantify.

Developing the data required for producing the field map is the most difficult, time-consuming, and technical part of precision farming. In general, spatially varying data are collected using two methods. First, data are measured directly from the field using sensors and instrumentation. Second, data are gathered from existing documentation in the library, such as United States Geological Survey (USGS) topographic maps, SCS (Soil Conservation Service) soil survey maps, and historical weather data from weather stations. All the data must be analyzed for error and any missing data estimated. Geostatistics methods have been used to estimate the missing data. University of Idaho personnel have used a geostatistics software package to estimate missing wheat yield data. The spatial dependence of wheat yield in the Palouse has been recognized for different field locations. Spatially variable mapping techniques have been widely used in recent research.

A 32.4-ha (80-acre) field near Steptoe, Washington was selected as the experimental site. The field has slopes of 0 to 45% and all aspects formed by a major ridge. This field lies in the Palouse region of eastern Washington, a region characterized by rolling to steep, dunelike, nonhomogeneous topography. The area topography has been classified as having 12 different landforms, and this classification was used to develop the concept of a typical Palouse hill. The typical hill has generally longer and less steep south- and west-facing slopes than in the north and east.

Real-time GPS positioning for the fertilizer applicator required use of on-the-go differential error corrections. A Trimble Community Base Station and Trimble Differential Global Positioning System (DGPS) mobile receiver were connected via an RTCM SC-104 radio link using Pacific Crest Instruments Radio modems. Trimble Navigation logging software was used with the base station. This equipment provided approximately 4.6-m (15-ft) accuracy in real time.

A 10-ft chisel plow was modified and used as the fertilizer applicator. A 757-l (200-gal) fiberglass tank was mounted on the hitch of the chisel plow and a peristaltic pump driven with a permanent magnet DC motor used to deliver fertilizer to individual tubes leading to the fertilizer shanks below the soil surface. A commercially available DC motor controller supplies the driving motor with a voltage proportional to a signal from the data acquisition card of the onboard computer. Output flow from the applicator is nearly linear with control signal level.

This project consisted of four parts:

1. Development of a procedure for determining the fertilizer rate for each cell;
2. Collecting the data for the specific field to be used in developing the database;
3. Development of equipment for applying fertilizer at a variable rate based on the database and determination of location; and
4. Field application of the fertilizer as required for each cell.

In the wheat field studied, the average yield varied from 4304 to 4710 kg/ha (64 to 70 bu/acre) in the previous 4-year period. For wheat production, the farmer applied a uniform nitrogen fertilizer rate of 72.9 kg/ha (65 lb/acre) for the whole field. As the sample consultation shows, the average nitrogen fertilizer rate predicted by the system was 44.8 kg/ha (40 lb/acre) for the average potential yield of 4708 kg/ha (70 bu/acre). Using the spatially variable fertilizer rate application system, the fertilizer rate was decreased by 25% of the rate the farmer would have used. An average nitrogen rate of 32.5 kg/ha (29 lb/acre) for the cells on 20% or greater slopes was used where the greatest potential of nitrogen leaching often occurs. The average nitrogen rate for cells on steep slopes was decreased by 54%.

Because the position is continuously updated (approximately once each second), no special machine path is required. The operator can work the field in a natural pattern and the GPS system and computer will locate the correct rate of fertilizer for the current cell. The operator could turn off the fertilizer by flipping a toggle switch in the tractor cab when overlapping previously fertilized areas.

Fertilizer Application—Application of fertilizer on the field ranged from 0 to 128 kg/ha (0 to 114 lb/acre) for individual cells. Control signals were recorded and checked against expected application rates, with excellent agreement.

7.8.2   VARIABLE VARIETY SEEDING EQUIPMENT

Precision farming equipment was also constructed to select from a choice of three small-grain varieties on-the-go according to a prepared map of the field (White et al., 1996). The ability to choose varieties might be used as follows: a farmer could place a cold-tolerant variety on the hilltops, a high-yielding variety on the slopes, and a lodge-resistant variety for the highly fertile bottomlands. The machine senses its location in the field via GPS, accesses the database to determine the desired variety, and selects that variety on-the-go through use of computer control and electric drive clutches on the grain-metering mechanism. In tests with the equipment during the spring of 1995, selection accuracy averaged 5.5 m (18 ft), the maximum error was 20 m (66 ft). No errors for selecting the varieties other than the edge effects were noted; i.e., no wrong variety was planted within the middle of a plot.

7.9 RECOMMENDATIONS

There are many types and styles of individual openers, press wheels, and cutting tools that may be combined to build complete conservation tillage planting machines. The investment required is significant and can have a major effect on the profitability of a farm operation. One considering adopting conservation seeding practices should read as much available literature for the local area as possible, consult with others who have tried or are using systems, ask for advice from farm advisors and county extension personnel, and, when possible, borrow or rent equipment for trial on the farm.

Morrison et al. (1989) developed an expert-system computer program called Planting, designed to systematically develop specifications for soil-engaging components on conservation planters, drills, and seeders. It is intended for use by farm advisors and others when consulting with a farmer on adoption of conservation tillage crop production. A report is printed that contains recommended planting machine components for each of the seven machine functions. They say that the required components could be added to existing machines for conservation planting.

Conservation seeding equipment has gone a long way in reducing erosion from fall-seeded wheat in the Pacific Northwest; continued research and development may yet someday reduce this soil loss well below the levels considered acceptable today.

REFERENCES

Haybuster Mfg., Inc., 1984. Haybuster 100 Series Drills, advertising literature, Haybuster Mfg., Inc., Jamestown, ND.

Hyde, G. M., J. E. George, K. E. Saxton, and J. B. Simpson, 1986. A slot-mulch implement design, *Trans. ASAE*, 29(1), 20–25.

Moden, W. L., Jr., 1965. Vertical Mulching in the Palouse Region, Idaho Agricultural Research Progress Report No. 110, University of Idaho, College of Agriculture, Moscow.

Morrison, J. E., R. R. Allen, D. E. Wilkins, G. M. Powell, R.D. Grisso, D. C. Erbach, L. P. Herndon, D. L. Murray, G. E. Formanek, D. L. Pfost, M. M. Herron, and D. J. Baumert, 1988. Conservation planter, drill and air-type seeder selection guideline, *Appl. Eng. Agric.*, 4(4), 300–309.

Morrison, J. E., Jr., S. H. Parker, C. A. Jones, R. R. Allen, D. E. Wilkins, G. M. Powell, R. Grisso, D. C. Erbach, L. P. Herndon, D. L. Murray, G. E. Formanek, D. L. Pfost, M. M. Herron, and D. J. Baumert, 1989. Expert system for selecting conservation planting machines: "PLANTING," *Trans. ASAE*, 32(2), 397–401.

Payton, D. M., G. M. Hyde, and J. B. Simpson, 1985. Equipment and methods for no-tillage wheat planting, *Trans. ASAE*, 28(5), 1419–1424 and 1429.

Peterson, C. L., E. A. Dowding, K. N. Hawley, and R. W. Harder, 1983. The Chisel-Planter minimum tillage system, *Trans. ASAE*, 26(2), 378–383 and 388.

Peterson, C. L., E. A. Dowding, K. N. Hawley, and J. C. Whitcraft, 1988. Shank and seed—a minimum tillage approach, in *Proceedings of the Seventh Annual Inland Empire Conservation Farming Conference*, Soil and Water Conservation Society, Moscow, ID.

Peterson, C. L., B. He, G. J. Shropshire, and K. B. Fisher, 1993. A spatially variable management system for the application of fertilizer for the production of winter wheat in the Palouse, *SAE Trans.*, 102(2), 312–326.

Pigeon, J. D., 1983. "Paraplow"—A New Approach to Soil Loosening, ASAE Paper No. 83-2136, ASAE, St. Joseph, MI.

Prasco. 1980. Prasco Super Seeder, LTD, advertising literature, Winnipeg, Manitoba, Canada.

Saxton, K. E., J. F. Kenny, G. M. Hyde, and L. F. Elliott, 1986. Slot mulch and Paraplow for conservation tillage, in *STEEP—Conservation Concepts and Accomplishments, Proceedings of the STEEP Symposium*, USDA ARS, Pullman, WA.

Swanson, G. 1983a. Yielder No-Till Drill, advertising literature, Yielder No Till Drill, Spokane, WA.

Swanson, G., 1983b. Yielder No-Till Drill, Tech Letter 83-1, Yielder No Till Drill, Spokane, WA.

Veseth, R., 1985. Research grain drill opener designs for conservation tillage, *PNW Conservation Tillage Handbook Series*, Chap. 2—Systems and equipment, STEEP extension conservation farming update, College of Agriculture, University of Idaho, Moscow.

Veseth, R., 1988. Conservation tillage considerations for cereals, *PNW Conservation Tillage Handbook Series*, Chap. 2—Systems and equipment, STEEP extension conservation farming update, College of Agriculture, University of Idaho, Moscow.

White, J. L., J. C. Whitcraft, J. C. Thompson, and C. L. Peterson, 1996. A Variable Variety Grain Drill for Wheat Production, ASAE Paper No. 961021, ASAE, St. Joseph, MI.

Wilkins, D. E., 1988. Apparatus for Placement of Fertilizer below Seed with Minimum Soil Disturbance, U.S. Patent 4,765,263.

Wilkins, D. E., 1996. Tillage, seeding and fertilizer application technologies, *Am. J. Alternative Agric.*, 11(2), 83–88.

Wilkins, D. E. and D. A. Haasch, 1990. Performance of a Deep Furrow Opener for Placement of Seed and Fertilizer, Columbia Basin Agricultural Research Special Report 860, Agricultural Experiment Station, Columbia Plateau Conservation Research Center, USDA ARS and Oregon State University, Pendleton.

Wilkins, D. E., G. A. Muilenburg. R. R. Almaras, and C. E. Johnson, 1983. Grain-drill opener effects on wheat emergence, *Trans. ASAE*, 26(3), 651–655 and 660.

Wilkins, D. E., P. E. Rasmussen, and J. M. Kraft, 1986. Effect of Paraplowing on Winter Wheat Growth, Columbia Basin Agricultural Research, Special Report 776, Agricultural Experiment Station, Oregon State University, Pendleton.

8 The Adoption of Soil Conservation Practices in the Palouse

John E. Carlson and Don A. Dillman

CONTENTS

8.1 INTRODUCTION

The Palouse and Camas prairies and the Columbia Plateau of Idaho, Washington and Oregon have been identified by the 1980 RCA appraisal as one of the four most severe soil erosion areas in the U.S. (Batie, 1983). The combination of topography, climate, soil characteristics and types of crops grown makes these areas extremely susceptible to soil erosion by water.

The severity of soil erosion in the region was recognized as early as the turn of the century. By the 1930s, activities to control the problem were under way; for example, the Palouse Soil Erosion Experiment Station was located at Pullman, Washington. Since that time, considerable efforts by federal and state government agencies, Washington State University and the University of Idaho have focused on the area's severe soil erosion problem. In 1975, a research program entitled, "Solutions to Economic and Environmental Problems (STEEP)," was initiated in Washington, Oregon and Idaho (Michalson and Papendick, 1991). This 15 year research and extension program brought together scientists from the three states and a variety of disciplines to work toward solving erosion problems in the Pacific

0-8493-1185-3/99/$0.00+$.50

Northwest. In addition, other private and public agency efforts to deal with erosion were intensified.

Have these research and extension efforts resulted in an increased use of erosion control practices? Have farmers become more aware of the erosion problem than they were 15 years ago? Has erosion decreased? What factors influence the adoption of soil erosion practices by farmers? Do other factors constrain the adoption of erosion control practices? The analysis presented here provides documentation of the changes that have taken place among farmers in the Palouse over the period from 1976 to 1990. In addition, it provides insights into the adoption process useful in understanding the factors affecting the adoption of new technology.

8.2   THE GEOGRAPHIC LOCALE[1]

Data for this analysis were collected from farmers and absentee landowners in the Palouse area of eastern Washington and northern Idaho over a period of 15 years.

Several characteristics make the Palouse region desirable for measuring the adoption of erosion control technology. First, agriculture in this region is homogeneous; virtually all farms grow some combination of wheat, barley, peas and/or lentils. Other crops have been tried over the time period of the study and some, such as canola, have increased in acres grown. Second, the technology of wheat farming has changed dramatically, providing a situation where innovativeness in farming can be readily measured. This setting also has severe erosion problems on virtually every farm, making it possible to assess which farmers are innovative in dealing with environmental concerns. Multiple factors contribute to the subtle but highly destructive erosion processes occurring in the Palouse. Slopes under cultivation are exceptionally steep, averaging 15 to 20% but ranging up to 50%. Soils are deep loess, windblown deposits that, despite severe erosion, continue to be productive. Rainfall averages 14 in./year at the western edge of the region increasing to 26 in. a few miles to the east where the Palouse hills meet the foothills of the Bitterroot Range of the Rocky Mountains.

Rain comes almost exclusively and gently in the winter and early spring, giving erosion a subtle character unlike the Midwestern United States where soil loss is dramatized by torrential spring downpours. Although rills and gullies appear in early spring, all traces of their existence can often be eliminated by one trip across the field with normal cultivation equipment. Larger gullies resulting from rainfall on frozen ground cause public concern, but the occasional nature of such major erosion hides the continuing substantial soil losses (Oldenstadt et al., 1982).

Until recent years most farm operators have been able to ignore erosion control without noticeable yield losses or other inconveniences (such as filling of roadside ditches and dirty streams). Substantial yield increases as a consequence of new crop varieties, pesticides, and fertilizers have concealed evidence that erosion is a problem. As a result, the adoption of erosion control practices has been slow and selective.

---

[1] This section has been adapted from Carlson and Dillman (1983).

8.3 THE SURVEYS

The STEEP research program has provided the impetus for several surveys in the Palouse area from 1976 to 1990. Collectively, they provide a longitudinal perspective on farmer attitudes and behaviors.

1976 Farmer Survey—The first survey was conducted in 1976 and asked a random sample of 306 Palouse farmers about their use of erosion control practices and attitudes toward soil erosion (Carlson et al., 1976). This survey provided benchmark data at the beginning of the STEEP program. The response rate for the Palouse survey was 92 percent.

1977 Absentee Landowner Survey—Absentee landowners were owners who lived outside of either Whitman or Latah County and who did not personally farm land they owned in these counties. A random sample of the owners, identified from tax records as owners with a mailing address outside of Whitman or Latah counties, were sent a mail questionnaire. The original sample included 384 names. Individuals who had recently sold their land and could not be located because of incorrect addresses or had management of the land transferred to a trustee, had died, or had failing health were dropped from the sample. For these reasons, the number of potential respondents declined to 317. A total of 206 completed and returned our questionnaire, a response rate of 65%.

1980 Re-survey of Farmers—In 1980 those farm operators who were surveyed in 1976 were re-interviewed. Of the original 306 farmers, 272 were re-surveyed. Losses were due to death, farm sales, and refusals. The 272 completed interviews represents an 89% response rate of those interviewed in the original sample.

1985 Farmer Surveys—Two samples were drawn for this study. A sample of no-till farmers was constructed with the help of SCS offices in both Whitman, County, Washington and Latah County, Idaho. A second sample of 151 names was drawn randomly from lists of all farm operators obtained from the ASCS offices in these counties. Names of 11 farmers who were no-tillers were removed from the random sample when it was desired to use the random sample as a control group of nonusers. Of the 187 no-till farmers, 174 interviews were completed, a 93% response rate. Of the 140 eligible control groups, 114 interviews were completed for an 81% response rate.

1990 Farmer Survey—The initial STEEP program was completed in 1990. To measure changes during the duration of the program another random sample of 246 Palouse area farmers was interviewed concerning attitudes toward soil erosion and use of conservation practices. The response rate was 76%. Farmers were asked the same erosion questions in both 1976 and 1990. Thus, these surveys provide measurements of change in conservation attitudes and in the use of erosion control methods.

The farm-operator surveys were based on random samples of about 25% of the total farm households in the Palouse. Data were collected by structured personal interviews ranging from 1 to 2 h in length.

8.4 THEORETICAL PERSPECTIVE

The adoption–diffusion model originally developed by Ryan and Gross in the 1940s and further developed by Rogers (1983) has perhaps been more utilized for practical application than any model in the social sciences. This model has been used to help understand the process by which significant and rapid changes in the United States have occurred. In addition, it has been used by change agents in cultures throughout the world.

Despite its popularity, the model has been severely criticized on several accounts (Warner, 1974; Rogers, 1976; Goss, 1979; Hooks, 1980; Brown, 1981). Nowak (1984:218) summarized the main criticisms.

> ...It was based on a socio-psychological orientation, it assumed a one-way, trickle-down communication process; it used a static and monolithic definition of innovation; it ignored the consequences of adoption; it emphasized individual resistance in explaining the failure to adopt; and it was largely dependent on survey research methodologies.

Despite criticisms, the model still stands at the forefront in terms of application whenever the diffusion of innovations is discussed. Its popularity alone suggests that the model has established an element of validity not often found among the social sciences.

The surveys in this project have focused on several issues related to the diffusion and adoption of soil erosion practices. Results suggest modifications to be incorporated into the adoption-diffusion model as presented by Rogers (1983).

The authors contend that three distinct factors must be addressed when trying to explain the diffusion of a particular innovation. These include, the *context* in which the diffusion occurs, the characteristics of the *innovation* itself and the characteristics of the *individuals* who adopt the innovation for their personal use (Dillman, 1985a).

The *context* includes factors in the sociophysical environment where the innovation occurs affecting the acceptance of an innovation but not directly connected to it. For example, the adoption of color TV sets was constrained by the number of color telecasts. As color telecasts, VCRs, and TV games became available, the speed of adoption of color TVs increased. The adoption of no-till in the Palouse was also influenced by the context. The high prices of fuel, and a strong emphasis on erosion control in the early 1970s, enhanced the adoption of no-till farming by encouraging farmers to look for ways of reducing costs and improving their farming practices.

Characteristics of the *innovation* are important factors in adoption of new technology. According to Rogers, innovations are adopted more rapidly if they are simple, have a definite advantage over other methods, if good results are immediately visible, if use of the technology is compatible with previously existing practices, if the practice can be tried incrementally, and if the necessary equipment is easily available (Rogers, 1983). Again, color TV is a good example in meeting most of the above criteria. The only disadvantage was that one either had to buy a whole set or not get one (i.e., the innovation was not divisible). The important lesson from the study of color TV which took more than 20 years after its development to reach 90% diffusion level was that its fundamental characteristics could not account for the slow rate of diffusion. No-till farming, on the other hand, does not have many of the characteristics that make it easily adoptable and, at the present time, it is not a common practice among Palouse farmers.

The *individual* is the third factor involved in the adoption of new technology. Much effort has been placed on this aspect of the adoption process, and often the blame for nonadoption of new technology has been placed on individuals to the exclusion of the other two factors. Past research has shown that few individuals are innovators. Innovators are anxious to try new ideas and adopt them once the ideas demonstrate their usefulness. At the other end of the spectrum are late adopters or laggards who wait until the innovation has proven useful for nearly everyone else before they "give it a try." Research has shown that the earliest adopters are people who are venturesome, tend to take risks, are highly educated, have higher incomes, and have the ability to comprehend abstract ideas (Rogers, 1983). Informal observation suggests that initial purchasers of color TVs had greater education and higher incomes. Similar characteristics have been shown to be correlated with increased adoption of erosion control practices.

It would be futile to look only at the personal characteristics of an individual— education, attitudes, age, etc.—in an effort to explain the use of control practices meaningfully. At the same time, characteristics of the innovation itself, e.g., the efficiency of a particular conservation practice, or of the adoption context, e.g., the influence of the current national farm subsidy policy, require the work of professionals in other disciplines. Efforts to incorporate all three were aimed at making sure the relevant concepts pertaining to the innovation and context were included in the authors' research. For example, the relationship between landlord and farmer forms a context that may influence individual behavior. Also, the complexity of no-till drills, an innovation with much erosion control potential, suggested that individual characteristics, e.g., mechanical skills, would be an important variable for handling such a complex innovation. Thus, this research was guided by an effort to focus on concepts that captured aspects of the innovation and adoption context, as well as those of a more individual nature.

8.5 THE SOCIAL CONTEXT OF EROSION CONTROL IN THE PALOUSE

As the authors began to gather information on farmers in the Palouse regarding their adoption of erosion control practices, it became evident that contextual factors might play a significant role in the rate of adoption of these practices. During the first interviews in 1976, the operators indicated that landlords hindered them from carrying out desired erosion control on leased land. Another common response blamed government programs for their inability to control soil erosion. Several other social context factors that might be important were also identified. Subsequent analyses focused on landlord–lessee arrangements, kinship arrangements, social background characteristics, government programs, and the networking arrangements among farmers. This section will report the results of the social context factors that were studied.

8.5.1 Landlord-Lessee Arrangements[2]

Absentee landlords in Whitman and Latah Counties were mostly older people who inherited their land or became owners through kinship ties. Not much land was owned by outside investors with no direct ties to the Palouse region. Fifty-seven percent of the absentee landowners were women and the average age was 63 years. They were highly educated (50% have college degrees) and had high incomes (59% have over $20,000 per year). About a third of their income was from their farmland.

Two-thirds of the respondents inherited land from parents or a spouse, and 14% bought land from relatives. One-fifth had bought land from someone other than a relative. Nearly 30% reported they (or their spouse) had once farmed the land. Thus, most landowners were tied to the Palouse through kinship or past residence. The average quantity of farmland owned was 324 acres, and most of the landowners (86%) leased all land to one farmer. Only 2% leased to more than two farmers. More than one-fourth (28%) of the respondents were not sole owners of their land. Instead they own an undivided interest (part ownership of an undivided farm), shared with 1 to 12 other people. There is some evidence that much of this undivided property is owned in part by the farm operator. Forty-two percent of the landowners said that major decisions about the land were made by another owner who was also the farm operator. Nearly 90% of the absentee owned land in the Palouse was under some form of crop-share agreement. About 50% of these agreements were written and half were oral agreements. They were also long term agreements with about a quarter being one year and the same number being 11 or more years. Another indication of stability was that 60% were with the current farm operator. About three quarters had never changed the farm operator.

It was concluded that absentee landowners did not impose major obstacles to erosion control. They tended *not* to be outside investors seeking maximum short-run returns at the expense of long-term productivity losses. Rather, they were tied to the Palouse by kinship and former residence as well as current landownership. Their values likely reflected those held by current residents of the region, as opposed to an outsider who might be more likely to hold different value orientations.

If absentee landowners were not a major obstacle to erosion control, neither did they seem to offer much potential as promoters of more conservation. Most land-owners were older women, many of whom were widows. They had little contact with farm operators and trusted them to make the "right" decisions. Information programs directed toward landowners might improve the climate for erosion control, but direct approaches to the farm operators would probably have more influence on erosion control decisions.

No support for the belief that landlords overtly deny carrying out erosion control practices on their land was found. Farm operators had a great deal of leeway in terms of initiating erosion control practices on both their land and the land they lease. A major finding from this work was the lack of communication among farm operators and absentee landlords regarding erosion control (Table 8.1).

---

[2] The results of the analysis of landlord–lessee arrangements are reported in detail in two articles (Dillman et al. 1978; Dillman and Carlson, 1982).

TABLE 8.1

Communication About Soil Erosion Between Landlords and Farm Operators as Seen by Farmers and Landlords

	Farm Operator Perspective				
Local Landlords	Absentee	Absentee Landlords			
Type of Communication	Landowner Perspective (N = 206)	Primary Landlord (N = 92)	Secondary Landlord (N = 64)	Primary Landlord (N = 125)	Secondary Landlord (N = 82)
	Percentages				
Talked to farm operator by telephone:					
Monthly or more frequent	26	39	19	67	64
A few times/year or less	74	61	81	33	36
Discussed soil erosion concerns with farm operator/landlord:					
Semi-annually or more frequent	8	13	10	41	33
Less than semi-annually	92	87	90	59	67

Neither had landlords abdicated their stewardship responsibility by establishing lease arrangements that encouraged exploitation by farm operators. If absentee landowners are having a significant effect on erosion control, the authors suggest that it may be by providing "a convenient excuse." The data neither proved nor disproved its existence so it remains a question for further research. Having cast doubt on the existence of other types of influence, the convenient excuse remains the most viable. It seems the absentee landlord was simply not as important a factor in erosion control as earlier surmised.

8.5.2 KINSHIP AND EROSION CONTROL[3]

Early in our research on soil erosion we observed that about a third of the farmers farmed with a relative. In most cases these were parent-child and sibling relationships but other combinations also existed. As the possible influences of these kinship relationships on erosion control were discussed two possibilities emerged. One would hinder adoption of new technology. Intergenerational (e.g., father–son) kinship ties provide cultural anchor points to older generations and established ways of doing things. These may repel pressures for change. Because of the control of the father, the young farmer may be constrained from letting old methods of farming give way to new methods.

It is also possible that kinship ties might encourage the adoption of new technology. The farmer who plans to sell the farm and retire rather than pass it on to the children may have a short-term planning perspective. Thus, the farmer may be hesitant to invest in long-term practices from which he or she will see little return. Also, more than one operator might provide greater opportunity for new information

---

[3] This section is a summary from material found in Carlson and Dillman (1983).

to come into the farm operation. It may also provide increased communication and discussion about what is best for the operation.

The results showed that "farming with a relative" has a significant positive effect on innovativeness in the area of soil erosion and a positive although not as strong effect on innovativeness in non-erosion-control areas. Some of this effect is direct and some is mediated through farm size and income level. In addition, the analysis indicated that farmers expecting one of their children to farm were more innovative in erosion control than those who do not have a child who is planning to farm. The data suggested that farmers were aware that erosion was something existing beyond one generation of farmers, whereas many other innovations did not have the same intergenerational implications. Not many farmers are concerned about their child inheriting a chisel plow; however, all farmers are aware of the likelihood of their son or daughter inheriting the land.

8.6   THE ROLE OF THE INNOVATION IN THE ADOPTION OF SOIL EROSION PRACTICES[4]

The innovation, or new technology, itself will affect its own adoption rate. Innovations tend to be adopted more rapidly if they are simple, if they have a definite advantage over other methods, if good results are immediately visible, if use of the technology is compatible with previously existing practices, if a little bit can be tried at one time and if the necessary equipment is easily available. Some innovations exhibit all these characteristics, whereas others may exhibit very few of them. The rate of adoption of soil erosion technologies in the Palouse reflects differences in the characteristics of the innovations themselves. For example, no-till farming was initiated in the Palouse in the early 1970s but has been adopted by fewer than a quarter of the farmers. On the other hand, the most commonly used control practice, seeding on the contour, is being used by close to 100% of the farmers. The difference in adoption is largely due to the characteristics of the innovation.

A no-till system contrasts sharply with seeding on the contour in terms of its characteristics. First, it is quite complex, requiring a farmer to understand the use of herbicides and fertilizer placement in a way that was not necessary for conventional tillage. Further, no-till has not always had a clear advantage over conventional tillage. In fact, most farm operators adopting no-till had to learn to "appreciate" the desirable qualities of a field with considerable residue on it. For many decades the quality of a field (and the farmer) was evaluated on the basis of the covering of all residue. The no-till system requires doing something quite different.

In addition, a no-till system requires many changes in farming methods, ranging from the timing of tillage practices to the kind of machinery that gets used. Although no-till is divisible, in that one can custom-hire a few acres done, and for the most part is available, the overall characteristics of the innovation suggest that it is likely to be adopted slowly. Figure 8.1 compares the characteristics of no-till with seeding on the contour.

---

[4] The material in the following section has been adapted from Carlson et al. (1987; 1994), Dillman et al. (1987).

FIGURE 8.1
A Comparison of Selected Characteristics of Seeding on the Contour with No-Till Farming

Seeding on the Contour	No-Till System
+ Simple	− Complex
+ Definite relative advantage	− Advantage not always clear
+ Immediately visible good results	− Must learn to appreciate results
+ Compatible with existing practices	− May require many changes
+ Divisible	+ May be divisible
+ Good availability	+ Good availability

Many of the changes in the use of erosion control practices in the Palouse over the past 15 years can be explained in part by the characteristics of the innovation itself (Carlson et al., 1994). Those easiest to use and showing the greatest differences in amount of erosion (i.e., minimum tillage, chisel plow, and divided slopes) have increased the most. Among eight practices traced over a 20–year period, no-till is still the least frequently used even though it has increased from about 3% in 1976 to 33% in 1995[5]. However, its rate of adoption into a farmer's ongoing farming practice has been much slower.

8.7   INDIVIDUAL FARMER CHARACTERISTICS AND EROSION CONTROL[6]

Of the three factors affecting the adoption of new technology—the context, the innovation, and the individual—blame for the lack of adoption of new technology is most often placed on the individuals. "If only farmers would realize how much better things would be if they only would use the computer, or the no-till drill, or the new-style combine, etc." This statement implies that farmers, themselves, are the primary constraint to the adoption of new technology. Yet, individuals are only one of three components that affect adoption of new technology. Research has shown, however, that individuals do vary in terms of the time they take to adopt new technology. This variation is influenced by a number of individual background characteristics such as age, amount of education, and the unique skills a person may have relevant to a given innovation. This was especially true of Palouse farmers in the mid-1970s, when age and education level were significant predictors of the number of conservation practices being used. Older farmers and farmers with higher education levels were more likely to use more erosion control practices than younger farmers and those with lower educational levels. However, the important factors affecting number of practices used has changed in the Palouse over the past 15 years (Table 8.2). In 1990, attitudinal variables about conservation in general and erosion

---

[5] Unpublished data from a STEEP survey conducted in 1995.
[6] Most of the material in this section is adapted from Carlson and Dillman (1988) and Carlson, Schnabel, Beus and Dillman (1994).

TABLE 8.2
Multiple Classification Analysis of Factors Influencing the Number of Erosion Control Practices Being Used in 1976 and 1990 in the Palouse

Variable	Background only 1976	Background only 1990	Background and Rainfall Zone 1976	Background and Rainfall Zone 1990	Background and Rainfall Zone and Erosion Problem 1976	Background and Rainfall Zone and Erosion Problem 1990	Background and Rainfall Zone and Erosion Problem and Conservation Attitude 1976	Background and Rainfall Zone and Erosion Problem and Conservation Attitude 1990
Age	.11	−.03	.11	−.02	.12	−.04	.11	−.06
Education	.16[a]	−.10	.16[a]	−.08	.16[a]	−.08	.17[b]	−.12
Size of Farm	.25[b]	.09	.26[c]	.12	.25[b]	.11	.25[b]	.12
Income	.05	.27[b]	.05	.20	.04	.21[a]	.04	.21[a]
Farm w/relative	.18[b]	.10	.18[b]	.09	.17[b]	.08	.17[b]	.10
Rainfall Zone			.07	.23[b]	.08	.25[c]	.07	.26[c]
Erosion Control					.13	.22[b]	.14	.13
Conservation Attitude							.07	.21[b]
R^2	.16	.12	.16	.16	.18	.21	.18	.24

[a] $p < .05$
[b] $p < .01$
[c] $p < .001$

control in particular were better predictors of number of erosion control practices used than either age or educational level. Two other variables commonly attributed to individuals are size of farm and income levels. Income level continued to be a statistically significant influence during the 15–year period. Size of farm was a significant predictor in 1976, but its influence diminished in 1990. In addition, research showed that the mechanical skill of the farmers was the most important factor associated with the adoption of no-till in the Palouse (Carlson and Dillman, 1988). It is important to consider the background characteristics of potential adopters of a new technology.

8.7.1 FARMER NETWORKS AND EROSION CONTROL[7]

Past research has shown that innovative farmers rely on mass media more frequently than on other farmers for new farming ideas. However, the findings suggest that farmers relied heavily on other farmers for information regarding innovative erosion control practices at all levels of the adoption process. As a group, no-till users cited other farmers as a source of no-till information more often than any other individuals (Table 8.3). (Dillman et al., 1987).

These data suggest that farmers were as important, or more important, than mass media or outside agencies in influencing the first farmers to use no-till (Table 8.4). This

---

[7] Most of the material in this section is adapted from Beck (1991).

TABLE 8.3
Occupation of the Most Important Person(s) Influencing Decisions to Use No-Till

Occupation	Users by Year of First Use			
	Group 1 1970–74 (N = 8)	Group 2 1975–78 (N = 32)	Group 3 1979–82 (N = 87)	Group 4 1983–84 (N = 45)
	Percentages			
No-till Farmer	75.0	81.2	49.4	75.5
Soil Conservation Service/Extension Agent	25.0	15.6	39.0	37.8
Chemical/Equipment Dealer	—	9.4	13.8	6.7
Other Farmers	—	—	2.3	11.1
Washington State University/University Employees	—	—	4.6	—

TABLE 8.4
The Single Most Important Influence on Farmers' Decisions to Use No-Till

Information Souce	Users by Year of First Use			
	Group 1 1970–74 (N = 8)	Group 2 1975–78 (N = 32)	Group 3 1979–82 (N = 87)	Group 4 1983–84 (N = 45)
	Percentages			
Local Farmer	40.0	43.3	28.6	40.5
Farm Magazines/Journals	40.0	—	2.6	—
Soil Conservation Service	20.0	23.3	39.0	42.9
Tour of No-till Fields	—	20.0	15.6	14.3
Machinery & Chemical Dealers	—	10.0	6.5	—
County Extension Office	—	3.3	2.6	—
Washington State University/University of Idaho	—	—	5.2	2.4

finding differs from past research showing that earliest users turn to outside or more "cosmopolite" information sources for new ideas (Wilkening, 1950; Rogers, 1983).

Our society has changed from a "community control" era where most of a persons needs could be met and their lives dominated by their local environment through a "mass society" era where standardization dominated and people looked to the federal government and national corporations for help to the present "information age" era where the focus is on people's abilities to organize, store, retrieve and transmit information (Dillman, 1985b). An important differentiation between mass society and the information age is access to information as it relates to hierarchy or social stratification. The information age reduces the importance of hierarchy, because everyone can get information at the same time, if they want to. The information age involves many workable alternatives for accomplishing different tasks, and what fits one enterprise or farming operation, may not fit another. The mass society was a time when it was not difficult to agree upon what farmer's "should" be adopting, and the pace of change was slower so that the "half-life" of a good technological development was longer (Allen and Dillman, 1994).

In the information age, with its lack of hierarchy, distinctions regarding availability of selected information sources at different stages in the adoption process may not be appropriate. The "laggard" may have a TV, VCR, satellite dish, and computer and be receiving information about new technology as rapidly as those who tend to be "innovators" or "early adopters." The defining characteristic of the information age is the increased tendency of individuals to seek out and use information without going through a hierarchy to get it. We should not be surprised that adoption no longer takes place in a neat hierarchy of scientist to innovator, to early adopter, to the early majority and late majority and finally, the laggards as suggested by the traditional adoption-diffusion model.

These findings also raise questions about the identification and importance of opinion leaders in the adoption process. If farmers are viewed as the best source about adopting a new technology, the role of opinion leaders may be enhanced and their selection more important in promoting adoption at all stages of the process. They may be perceived as a more stable and trustworthy source of information in a society where information overload is common. In fact, their identification as opinion leaders may be based more on their ability to filter and transmit appropriate information than on having certain social background characteristics.

If the information age reduces the hierarchical aspects of society, it means that farmers are less likely to adopt new innovations on the basis of well established social characteristics, such as age, income, size of farm, etc. but rather on the basis of their access to particular kinds of information. Beck (1991) tested this in his analysis of the early adopters of no-till by utilizing two network variables in addition to the more traditional background variables. With the two network variables Beck measured the effect of a hierarchical approach common in the mass society with a non-hierarchical approach more consistent with the information age.

The hierarchical approach suggests that individuals are influenced the most by the people with whom they directly interact. Strong research support for this model as a factor in the diffusion of innovations has been shown in a number of areas (Coleman, Katz and Menzel, 1966). This model is more consistent with the mass society assumptions that information is passed down through channels to the final adopter. Those who interact together are most likely to share common background characteristics; thus, those who adopt early should differ on certain background characteristics from those who adopt later.

The non-hierarchical approach differs from the hierarchical by hypothesizing that individuals are most influenced by the people they perceive as being similar to themselves, regardless of how often they interact with them (Burt, 1987). It is not personal interaction but perceptions based on information about others who are similar to themselves. In an information age this approach should do a better job of predicting adoption than both the cohesion model and background characteristics.

Analysis of data on the adoption of no-till in the Palouse showed that farmers followed a non-hierarchical pattern to a greater degree than a hierarchical pattern.

The only background variable to correlate significantly with year of first use was age, with older farmers more likely to adopt first (Beck, 1991).

The analysis suggests that it may no longer be valid to differentiate farmers as being an innovator, early majority, late majority and laggards on the basis of background characteristics. It is also likely that these categories do not apply in all areas of a farmers life. A dairy farmer may have the most up-to-date feeding system but be a laggard when it comes to his haying machinery.

CONCLUSION

Society is much different today than it was when the adoption–diffusion model was originally developed. Dillman and Beck (1986) have aptly discussed these changes in terms of a move from a community control era, through a mass society era and now toward an information age era.

These changes have important ramifications about how information about new technology is transferred to potential users and who farmers perceive as being important referents in the adoption process. Since "communication through certain channels (Rogers, 1983, p. 10)" is a major part of the diffusion process, social changes in the information area are significant for diffusion. Dillman (1985b) identified the major changes associated with a move from mass society to the information age. He discussed changes that included increased speed, amount, and ability to receive and send information worldwide.

The adoption model was fitted to the mass society, with its emphasis on hierarchy, and the multistep flow of information. The diffusion pattern in the mass society was primarily a trickle-down of information from those most closely in touch with the original sources of new technology, eventually ending up with the laggards. Different kinds of information were shown to be most influential in the adoption process at various stages in a person's decision-making process.

The Palouse research produced several findings that suggest that the traditional model needs modification to fit the information age. The discussion of the diffusion process as involving the *context*, the *innovation*, and the *individual* provides a framework for identifying the various components of the process into three categories that help explain the adoption process. The findings suggest that the information age might influence how we categorize individuals. The common notion of innovators, early adopters, early majority, late majority and laggards may no longer be important in predicting the flow of diffusion. Individuals appear to be the most important influencers of adoption at all stages of the process. Farmers were more likely to relate with other farmers in developing and adopting new innovations (no-till drills in our research) instead of looking to a university or industry researcher for new ideas.

Kinship and network relationships were also shown to be important in adopting new technology. However, networks were based more on communication with persons of similar backgrounds rather than on the basis of frequency of personal contact. Those farmers who farmed with a relative were more conservation oriented than farmers who farmed alone. It is likely that those who farmed with a relative would have a longer term perspective than those who farmed alone. They also would be more likely to have someone to discuss the issues related to soil conservation with.

There is little chance that the adoption–diffusion model will be replaced in its entirety by a new theoretical paradigm. In a general sense, the model has been useful in guiding the diffusion of innovations for many years. Perhaps its success has hindered its further development in the sense that it is difficult to suggest that a model as useful as the adoption model might need additional theoretical development. This is not to say that the model has remained entirely static. Rogers (1983) has continued to update the model, attempting to deal with criticisms and changes. Yet people do not live in the same society that existed at the time of the model's development. Continued research and theory development on the adoption and diffusion of new technology is important to improve a model that has been shown to be one of the most popular and useful models in the social sciences. If we are going to understand adoption in the information age, when things happen so very fast, we absolutely must watch and understand how to analyze the context and its relationship to both innovations and individuals.

REFERENCES

Allen, J. and D.A. Dillman, 1994. *Against All Odds*, Westview Press, Boulder, CO.

Batie, S., 1983. *Soil Erosion: Crises in America's Croplands*, The Conservation Foundation, Washington, D.C.

Beck, D.M., 1991. The Impact of Social Network Variables on the Diffusion of No-Till Agriculture, Ph.D. dissertation, Department of Sociology, Washington State University, Pullman.

Brown, L.A., 1981. *Innovation Diffusion: A New Perspective*, Mathuen, London.

Burt, R., 1987. Social contagion and innovation: cohesion versus structural equivalence, *Am. J. Sociol.*, 92, 1287–1335.

Carlson, J.E. and D.A. Dillman, 1983. Influence of kinship arrangements on farmer innovativeness, *Rural Sociol.*, 48(2), 183–200.

Carlson, J.E. and D.A. Dillman, 1988. The influence of farmers' mechanical skill on the development and adoption of a new agricultural practice, *Rural Sociol.*, 53(2), 235–245.

Carlson, J.E., M.E. McLeod, and D.A. Dillman, 1976. Farmers' Attitudes toward Soil Erosion and Related Farm Problems in the Palouse Area of Northern Idaho and Eastern Washington, Progress Report No. 196, Agricultural Experiment Station, University of Idaho, Moscow.

Carlson, J.E., D.A. Dillman, and C.E. Lamiman, 1987. The Present and Future Use of No-Till in the Palouse, Research Bulletin No. 140, Agricultural Experiment Station, University of Idaho, Moscow.

Carlson, J.E., B. Schnabel, C.E. Beus, and D.A. Dillman, 1994. Changes in the soil conservation attitudes and behaviors of farmers in the Palouse and Camas prairies: 1976–1990, *J. Soil Water Conserv.*, 49, 493–500.

Coleman, J., E. Katz, and H. Menzel, 1966. *Medical Innovation*, Bobbs-Merrill, New York.

Dillman, D.A., 1985. Factors influencing the adoption of no-till agriculture, in D. Huggins, Ed., *Proceedings, 1985 No-Till Farming Crop Production Seminar*, Yielder Drill Company, Spokane, WA.

Dillman, D.A. and J.E. Carlson, 1982. Influence of absentee landlords on soil erosion control practices, *J. Soil Water Conserv.*, 37, 37–41.

Dillman, D.A., J.E. Carlson, and W.R. Lassey, 1978. The Influence of Absentee Landowners on Use of Erosion Control Practices by Palouse Farmers, College of Agriculture Research Center, Washington State University, Pullman, Circular 607.

Dillman, D.A., D.M. Beck, and J.E. Carlson, 1987. Factors influencing the diffusion of no-till agriculture in the Pacific Northwest, in L. Elliott, Ed., *STEEP-Soil Conservation Concepts and Accomplishments*, Washington State University Press, Pullman.

Goss, K., 1979. Consequences of diffusion of innovations, *Rural Sociol.*, 44, 754–772.

Hooks, G., 1980. The classical diffusion paradigm in crisis, paper presented at the annual meeting of the Rural Sociological Society, Ithaca, NY (unpublished).

Michalson, E. and B. Papendick, 1991. STEEP: a regional model for environmental research and education, *J. Soil Water Conserv.*, 46(4), 245–250.

Nowak, P., 1984. Adoption and diffusion of soil and water conservation practices, in English, B., J. Maetzold, B. Holding, and E. Heady, Eds., *Future Agricultural Technology and Resource Conservation*, Iowa State University Press, Ames.

Oldenstad, D.L., R.E. Allen, G.W. Bruehl, D.A. Dillman, E.L. Michalson, R.L. Papendick, and D.J. Rydrych, 1982. Solutions to environmental and economic problems: STEEP, *Science*, 217 (3 Sept.), 904–909.

Rogers, E.M., 1976. Communication and development: the passing of the dominant paradigm, in E. Rogers, Ed., *Communication and Development: Critical Perspectives*, Sage Publications, Beverly Hills.

Rogers, E.M., 1983. *The Diffusion of Innovations*, The Free Press, New York.

Warner, W.K., 1974. Rural sociology in a post-industrial agriculture, *Rural Sociol.*, 39, 306–318.

Wilkening, E., 1950. Sources of information for improved farm practices, *Rural Sociol.*, 15, 19–29.

9 A Systems Approach to Conservation Farming

Douglas L. Young, Frank L. Young,
John E. Hammel, and Roger J. Veseth

CONTENTS

9.1   INTRODUCTION

Conservation farming systems are a subset of general cropping systems. Conservation farming systems add the objectives of soil, water, and air quality protection to those of achieving acceptable crop yields and adequate farm income.

The landmark 1985 Farm Bill added urgency to the drive to develop and implement effective conservation farming systems. Under the Conservation Compliance provision of this legislation, farmers with highly erodible land were required to file approved soil-conserving farm plans by 1990 and fully implement them by 1995 to remain eligible for USDA programs and payments. These programs include the financially important 1996 Farm Bill transition payments, federal crop insurance, and Farmers' Home Administration loans, among others. The 1985 legislation also initiated the Conservation Reserve Program (CRP) which paid farmers to retire erodible land to perennial vegetative cover for 10 years. Over 2 million acres were entered into the popular CRP in Washington, Oregon, and Idaho. Some of these erodible lands began emerging from contract in 1996, but growers need to comply with acceptable conservation compliance systems if they are to return CRP land to production and retain program eligibility.

Another factor that has promoted conservation systems has been the upsurge in public interest regarding agricultural sustainability. The public interest in environmentally sound and sustainable farming means that conservation is likely to remain important regardless of the durability of conservation provisions in the 5-year cycle of federal farm legislation. Federal, state, and local environmental agencies are scrutinizing the impact of farm production practices on soil, water, and air quality.

Finally, economic factors are inducing the adoption of conservation farming. As management technologies improve and grower experience with conservation farming increases, economic benefits are emerging. These include the potential for higher yields where surface residues increase water conservation, and lower production costs, primarily from fewer or combined field operations.

Successful conservation farming systems involve integrating several of the components discussed in other chapters. These include use of suitable crop rotations, nutrient management, pest control, and tillage and residue management. Other chapters in this book provide detail on design and adaptation of pest management, nutrient management, and other components of farming systems for varying agroecological zones. Management practices are situation specific, and each grower will have to tailor these components to a greater or lesser degree to accommodate the farm environment. To assist growers in this design process, this chapter traces the evolution of a successful conservation farming system for one particular farming region—the $1/2$ million acre annual cropping region of the eastern Palouse located in eastern Washington and northern Idaho.

The chapter describes selected research trials in the subregion, which led to the identification of an economically and agronomically successful system. The research occurred over a two decade period at the University of Idaho, Washington State University, and at USDA-ARS research sites in the subregion. The learning process accompanying this research highlights the importance of a "full systems approach" in identifying successful conservation farming systems. This history illustrates that a successful conservation farming system is "only as strong as its weakest link."

9.2   COMPONENTS OF A SUCCESSFUL CONSERVATION FARMING SYSTEM

As a framework for evaluating progress toward successful conservation farming systems, it is useful to enumerate the characteristics of such a system. While the relative importance of different characteristics will vary from region to region and from farmer to farmer, each of these characteristics is essential for a complete conservation farming system.

TABLE 9.1
Most Profitable Weed Management Levels by Crop and Tillage,
USDA IPM Experiment, Pullman, Washington, 1986–91

| | Tillage | |
Crop	Conservation	Conventional
Winter wheat after winter wheat	MOD	MIN
Winter wheat after spring wheat	MOD	MIN
Winter wheat after spring peas	MAX	MOD
Spring wheat	MOD	MIN
Spring barley	MAX	MOD
Spring peas	MOD	MOD

Weed Management Level: MIN = minimum, MOD = moderate, MAX = maximum.

Source: Young, D. L. et al., *J. Soil Water Conserv.*, 49, 581–586, 1994. With permission.

Acceptable conservation farming systems should accomplish the following:

1. **Provide adequate disease, weed, and insect control.** Higher levels of surface residue and lower soil temperatures associated with conservation farming systems can increase crop losses from root diseases unless growers utilize management options for disease control. Mechanical weed control is limited in conservation systems, which depend upon reduced tillage and increased surface residue. Consequently, chemical or other control practices generally must be increased to substitute for the loss of mechanical weed control. Conversion to conservation tillage for a long-term cropping trial in the Palouse generated a general increase in the economically justified level of chemical weed control (Table 9.1). Some insect pests can be a problem in conservation tillage systems with continuous cereals or other short rotations. Cook and Veseth (1991) emphasize that crop rotation is the most effective way to combat root and other soilborne diseases, manage weeds, especially grassy winter annuals, and control insect pests. Other important management considerations include seeding date, fertilizer placement, varietal selections, and early control of volunteer cereals and weeds between crops.

2. **Maintain stand establishment and survival.** Under reduced tillage, effective planting equipment is necessary to ensure crop germination and emergence. Seeding drills must adequately penetrate residue and soil to ensure good seed-to-soil contact. This is of special concern in the high-residue, annual cropping areas of the eastern Palouse. One potential benefit of conservation tillage is increased overwinter survival of winter wheat. More snow is retained on fields with increased residue and surface roughness, providing greater insulation of the crop beneath.

3. **Supply crop nutrient needs.** Fertilizer timing and placement are especially critical under no-till and minimum-tillage systems. Adequate nutrient availability helps to minimize environmental stress and optimize yield potential.

4. **Maintain or improve soil moisture supply.** Soil moisture is the major yield-limiting factor in the dryland cropping areas of the Pacific Northwest. Properly managed conservation tillage can increase water infiltration into the soil and reduce losses by runoff and evaporation. However, field operations on wet soil should be avoided in all tillage systems to prevent compaction, reduced infiltration, restricted root growth, and poor water uptake.

5. **Maintain reasonable production costs.** In many regions, conservation tillage and conventional tillage costs are similar, but these depend upon the costs of planting systems, fuel costs, and other variable factors. Structural conservation practices such as terraces and grass waterways incur substantial up-front investments.

6. **Provide a profitable and stable income flow.** Rotational adjustments to achieve conservation objectives that decrease the frequency of harvested crops or include a switch from more profitable to less profitable crops can reduce annual incomes. The ability of conventional vs. conservation tillage systems to sustain yields during drought, adverse winter weather, and other stressful conditions will influence the stability of farm income.

7. **Ensure compatibility with government commodity programs.** Prior to the introduction of decoupled transition payments in the 1996 Farm Bill, participating growers had to confine wheat and barley plantings to permitted amounts of their historical wheat and barley acreages. Conservation cropping systems that violated these acreage restrictions were ineligible for USDA farm payments. This ineligibility often made such systems financially infeasible. If acreage restrictions return in future USDA programs or in regional marketing orders for selected crops, commodity programs could again influence the design of economically viable conservation cropping systems.

8. **Achieve soil conservation, water quality, and air quality targets specified by federal, state, and local agencies.** Eligibility for federal farm programs, including the 1996 Farm Bill transition payments to grain growers, is contingent upon full compliance with soil-conserving plans for erodible farmland. In many regions, state environmental agencies are also becoming increasingly involved in regulating the impact of agriculture on resource quality.

9.3 CONSERVATION TILLAGE RESEARCH IN THE EASTERN PALOUSE

Two long-term research trials are closely examined in this chapter to provide perspective on the evolution of conservation farming systems. These two experiments, the USDA-Integrated Pest Management (IPM) trials near Pullman, Washington

(1986 through 1991), and the University of Idaho (UI) Solutions to Economic and Environmental Problems (STEEP) trials near Moscow, Idaho (1974 through 1987), were selected by virtue of their long duration and systems approach. The trials simultaneously evaluated crop rotations, tillage, and selected cultural practices. Other, similar research trials in this subregion lasted only 1 to 3 years. Short-term research provides limited information for assessing the durability of a complex system. Multiyear research is needed to move beyond the agronomic transition phase and to assess accurately the impact of a system on economic risk by encompassing diverse weather conditions and several rotational cycles.

Both the recent IPM and the earlier STEEP trials were conducted in the 18 to 24 in. annual cropping region of the eastern Palouse. Long-run average precipitation for Pullman, Washington is 21 in./year and 22 in. in Moscow, Idaho. Palouse, Thatuna, and Naff silt loam soils dominate in this region. Typical are 2-year and 3-year rotations of small grains with dry peas and lentils. Winter wheat yields of up to 100 bu/acre in this region exceed those of any other major semiarid dryland production area in the U.S. The eastern Palouse is intensely cultivated and contains relatively few livestock. Farms are large, averaging 1489 acres in eastern Whitman County, Washington and 965 acres in western Latah County, Idaho (Halvorson, 1991). Slightly less than one half of the farm acreage was in wheat base and about one fifth was in barley base in 1990.

9.3.1 Description: University of Idaho STEEP Trials

This research was initiated in 1974 with funding provided by the USDA STEEP program. Dr. Roger Harder and then Dr. John Hammel of the UI Plant and Soil Sciences Department supervised the project until it was terminated in 1987 (Harder et al., 1980; Hammel, 1989, 1993). The field experiment, established 1 mile north of Moscow, Idaho, was situated on a south-facing hillside with 10 to 20% slope. Predominant soils at the site were Palouse and Naff silt loams, two similar soil series. In the experimental design, the main plot treatments consisted of combinations of three tillage systems and three crop rotations. Prior to 1982, four wheat varieties were used in subplots.

The conventional tillage treatment was essentially a moldboard plow, disk, harrow, and plant sequence. The minimum-tillage treatment involved a single pass with a UI-designed combination chisel-planter (Peterson et al., 1979), a combination chisel plow and double-disk planter. The chisel-planter deep-banded fertilizer below the seed. No-till seeding was accomplished with a John Deere Power-till drill modified to apply starter fertilizer with the seed. Primary tillage depths for the conventional and minimum-tillage treatments were 8 and 4 in., respectively. The following crop rotations were used: winter wheat/spring pea (2-year); winter wheat/spring barley/spring pea (3-year); winter wheat/spring pea seeded with alfalfa + red clover/alfalfa + red clover (3-year). There were four replications of each tillage-rotation combination on plots measuring 50 by 100 ft. Crops in each stage of the rotation sequence were grown each year.

After 1982, minimum tillage included fall planting of winter wheat using a chisel and drill sequence, whereas no-till involved the direct-drilling of fall and spring crops (Hammel, 1989). In addition, the 3-year rotation with alfalfa and red clover was dropped.

A base rate of 80 lb/acre of nitrogen (N) was applied to all wheat plots prior to 1982. Thereafter, a fall application of 100 lb N per acre was used without any subsequent spring top-dress application. Fertilizer was broadcast and incorporated under conventional tillage management, whereas fertilizer was shank-applied under both minimum and no-till systems.

Weed control on conventional tillage and minimum-tillage winter wheat included use of broadleaf and wild oat herbicides in the spring. No-till plots sometimes received preplant glyphosate herbicide treatments and postemergence applications of metribuzin in the spring to control downy brome. Downy brome infestations became severe in both the minimum and no-till treatments between 1974 and 1982. Aggressive, intensive weed control was utilized between 1983 and 1987 to remove weeds as a variable in the study.

9.3.2   DESCRIPTION: USDA IPM TRIALS

The USDA IPM research was initiated in the fall of 1985 under the supervision of USDA research agronomist, Dr. Frank Young. More than a dozen scientists from the USDA, University of Idaho, and Washington State University contributed to the research.

The IPM site is located 4 miles north of Pullman, Washington on Palouse silt loam soils in a 21-in.-average annual precipitation area. Primary treatments consisted of combinations of two rotations (winter wheat/winter wheat/spring wheat and winter wheat/spring barley/spring peas), three weed management levels (MAX, MOD, MIN), and two tillage systems (conservation tillage and conventional tillage).

The MAX level of weed management was generally specified as the rate and mix of herbicides required to achieve maximum control. The MOD level represented typical cooperative extension management recommendations. The MIN level was often a single tank mixture of two herbicides for grass and broadleaf weeds. Over 6 years, the MOD and MIN levels of control averaged about three quarters and one half of the cost, respectively, of the MAX level of control. Herbicide treatments varied year by year within general guidelines for the MIN, MOD, and MAX levels. Treatments represented a combination of nonselective, preemergence, preplant-incorporated, and postemergence products (Boerboom et al., 1993). Maximum and moderate weed management levels were within label requirements; minimum levels were often below labeled rates.

Primary tillage in the conventional system consisted of moldboard plowing wheat and barley stubble and disking pea residue. The conservation tillage system was a mix that combined no-till and minimum till. The initial winter wheat crop in both rotations was seeded by a no-till drill, which placed the fertilizer 4 in. below the soil surface between paired rows. The succeeding two crops were planted using minimum tillage following fall chisel plowing. Both conservation and conventional tillage plots received secondary tillage generally consisting of two field cultivations. All winter wheat plots received average applications of 116 lb N per acre of nitrogen fertilizer.

9.4 RESULTS: CONSERVATION SYSTEMS RESEARCH

9.4.1 University of Idaho STEEP Trials Results

The UI STEEP trials, which were initiated in 1974, were among the earliest long-term research projects on conservation tillage systems in the Pacific Northwest annual cropping region. Consequently, these trials provided an important foundation for subsequent research.

No overall summary of the entire 14-year history of these trials was published because of treatment changes during this period. However, agronomic and economic analyses of portions of the results were released periodically, which provide a reasonable understanding of the results (Harder et al., 1980; Young et al., 1984; Hammel et al., 1987; Hammel, 1989; 1993).

Since winter wheat is the most economically important crop in the region, the impacts of alternative practices, such as conservation tillage, on wheat yields is of great concern to growers. Throughout the UI STEEP study, winter wheat yields tended to be highest under conventional tillage and lowest under no-till (Table 9.2). Minimum tillage yields were intermediate. The percent yield reduction under conservation tillage was similar in both the early and final stages of the study. Dry pea yields were also lower under conservation tillage (Table 9.3). During the 1979 through 1983 period, downy brome infestations became severe, seriously limiting winter wheat yields in both the minimum and no-till treatments.

Year-to-year yield variability for winter wheat, which is an important indicator of risk potential, was lowest for no-till, intermediate for minimum till, and highest for conventional tillage during the 1976 through 1979 period. There was little difference in yield variability among tillage systems from 1984 through 1987.

Crop rotation can also affect yields. From 1976 through 1979, there was no consistent effect of crop rotation on wheat yields (Harder et al., 1980). However, in the final years of the trials, winter wheat in the 3-year rotation always yielded more than wheat in the 2-year rotation (Table 9.4). This rotation benefit averaged 10%

TABLE 9.2
Influence of Tillage on Winter Wheat Yields, UI STEEP Plots, Moscow, Idaho

| | Average Wheat Yield | | | |
| | 1976–79 | | 1984–87 | |
Tillage	bu/ac	Index	bu/ac	Index
Conventional	73	100	99	100
Minimum	67	92	91	92
No-till	59	80	78	78

Yields averaged over rotation. Index expresses yield as a percent of conventional tillage yield.

Sources: Harder et al. (1980) and Hammel (1993). With permission.

TABLE 9.3
Influence of Tillage on Dry Pea Yields,
UI STEEP Plots, Moscow, Idaho

Tillage	Pea Yield 1976–81	
	Cwt/Ac	Index
Conventional	11.3	100
Minimum	9.4	83
No-till	10.2	90

Yields averaged over rotation. Index expresses yield
as a percent of conventional tillage yield.

Source: Harder et al. (1980) and Young et al. (1984).

TABLE 9.4
Influence of Crop Rotation on Winter Wheat Yields,
UI STEEP Plots, Moscow, Idaho, bu/acre

Rotation	1984	1985	1986	1987	Average
WW-SP	97	57	77	115	87
WW-SB-SP	103	60	90	123	94

Yields averaged over tillage. WW = winter wheat, SB = spring barley,
SP = spring peas.

Source: Hammel, J. E., unpublished, 1993.

over the 1984 through 1987 period. Crop rotation did not appear to affect wheat yield variability.

Researchers offered several possible explanations for the yield reductions under conservation tillage. Moisture and nitrogen shortages may have restricted the growth of no-till wheat (Harder et al., 1980). Higher soil bulk densities on the no-till plots appear to have impeded infiltration rates. Soil compaction and soil strength were greater on the minimum- and no-till plots (Hammel, 1989). Profile water extraction by winter wheat was also less under minimum- and no-till management (Hammel, 1993). Hammel concluded that lower yields in the conservation tillage treatments occurred as a result of reduced water extraction, possibly because of the soil compaction that slowed or hindered root growth. While disease could also have reduced yields, no obvious disease problem was observed (Harder et al., 1980). However, several root diseases, which were difficult to diagnose until late in this study, may have been a factor in yield reductions under conservation tillage.

Another factor that may have limited crop yields in the minimum- and no-till treatments was the type of planting equipment used. Seeding depth was variable and stands were often erratic. The fertilizer placement capabilities of the modified John Deere drill used were less sophisticated than those available on subsequent no-till drill designs, including that of the USDA no-till drill used on the IPM experiment.

During the 1974 through 1982 years of the study, part of the fertilizer was placed with the seed, while the remainder was surface broadcast. Concurrent STEEP research showed that surface application of nitrogen fertilizer contributed to weed problems and nitrogen tie-up by surface residues (Koehler et al., 1987). After 1983, fertilizer was either shanked in (minimum till) or banded (no-till). Therefore, the effects of drill configuration and fertilizer application were minimized.

The UI chisel-planter caused considerably less soil disturbance than most "minimum" tillage systems. Such low-intensity tillage systems may contribute to soil compaction problems, which affect crop root growth, thus decreasing crop water and nutrient uptake (Hammel, 1989).

The overall poor performance of the continuous no-till system in the UI STEEP plots reflected similar experiences of growers attempting to implement no-till farming. The problems encountered during this 14-year study helped shape future research and equipment design efforts that have made no-till a more viable practice in the annual cropping region.

9.4.2 OTHER EARLY CONSERVATION TILLAGE RESEARCH RESULTS

Although other conservation tillage research was completed in the 1970s and early 1980s, none equaled the longevity of the UI trials reported above. For example, USDA researchers at Pullman compared no-till under various stubble treatments to conventional tillage from 1977 through 1979 (Cochran et al., 1981). No-till winter wheat yields over the 3 years were similar to those from conventional tillage, but no inferences can be made about the long-term viability of no-till because new sites were used each year.

Spring barley and spring wheat yields were similar across tillage systems (conventional, minimum, no-till) in field plots at Dayton and Pullman, Washington during 1979 and 1980 (Reinertsen et al., 1983a, b). The lack of yield loss under conservation tillage for these spring crops was encouraging, but few conclusions can be drawn from these 1- or 2-year studies. Also, these studies did not examine the effect of conservation tillage within a complete rotation system.

9.4.3 SURVEY RESEARCH: FARMERS' YIELD PERCEPTIONS

Young et al. (1984) reported the perceptions of eastern Palouse farmers surveyed in 1980 regarding the yield performance of conservation tillage systems for winter wheat and dry peas (Table 9.5). At that time, growers expected yield losses from no-till and minimum tillage. This was consistent with the UI experimental results. Some 25% of the 167 growers surveyed expected no-till to reduce wheat yields by 15 bu/acre or more. Another 53% expected smaller losses. Only 4% of the responding farmers expected winter wheat yield gains with no-till. Over half of the surveyed growers expected no-till to cut pea yields by 600 lb/acre or more. In 1980, Palouse growers expected wheat and pea yields to fare better under minimum tillage than

TABLE 9.5
Frequency Distribution of Eastern Palouse
Farmers' Expected Changes in Average Yields
When Conservation Tillage is Adopted

Yield Change Relative to Conventional Tillage	Percent of Respondents	
	Min-Till	No-Till
Winter Wheat (bu/ac)		
>15 loss	5	25
1–15 loss	30	53
No change	60	18
1–10 gain	5	4
	$(n = 175)$	$(n = 167)$
Dry Peas (lb/ac)		
>600 loss	7	51
201–600 loss	29	39
1–200 loss	22	4
No change	38	4
1–300 gain	4	2
	$(n = 105)$	$(n = 91)$

Source: 1980 STEEP survey as reported in Young et al. (1984).

under no-till, but many growers still expected yield reductions under minimum tillage (Table 9.5).

9.4.4   USDA IPM Trials Results

9.4.4.1   Yield Comparisons

The 6-year USDA IPM conservation farming trials, initiated in the fall of 1985 near Pullman, Washington, represent the most current and most complete conservation farming systems research in the Palouse region. In addition to two tillage treatments and two crop rotations, three levels of weed management were included in the experimental design. Weed control had been a major limitation with conservation tillage in previous research and farm experience in the Palouse, so its inclusion was important. In addition, the USDA IPM trials made use of improved fertilizer placement technology for no-till drills. Plant pathologists and entomologists served on the interdisciplinary IPM team to assess problems and make recommendations regarding pest management.

Kwon (1993) compared the 6-year average yield performance of conservation and conventional tillage systems for the IPM "main" trial (Table 9.6). On the average, no-till winter wheat outyielded conventional tillage winter wheat by 11 to 12%. Average yields for minimum-till spring barley and spring peas exceeded those for conventional tillage by 6 and 7%, respectively. Second-year winter wheat and spring wheat in the monoculture wheat rotation represented the only exceptions to the

TABLE 9.6

Comparison of Average Crop Yields for Conservation and Conventional Farming Systems, USDA IPM Experiment, Pullman, Washington, 1986–91

Rotation/Crop/Tillage		Weed Management Level			Average Yield
		MIN	MOD	MAX	
WW-SB-SP Rotation					
			bu/ac		
WW	Conventional	91	96	98	95
	No-till	96	107	112	105
			ton/ac		
SB	Conventional	2.2	2.4	2.4	2.3
	Minimum	2.2	2.5	2.6	2.4
			cwt/ac		
SP	Conventional	17.5	18.6	18.8	18.3
	Minimum	17.8	20.1	20.9	19.6
WW-WW-SW Rotation					
			bu/ac		
WW1	Conventional	78	77	79	78
	No-till	81	89	90	87
WW2	Conventional	72	70	73	72
	Minimum	69	72	75	72
SW	Conventional	60	60	60	60
	Minimum	53	58	58	57

All yields were adjusted to typical moisture and chaff levels in the eastern Palouse: 10% for wheat, 7.5% for barley, and 5% for peas. WW = winter wheat, SW = spring wheat, SB = spring barley, SP = spring peas.

Source: Kwon, T., Ph.D. dissertation, Washington State University, Pullman, 1993.

superior performance of conservation tillage. Yields for second-year winter wheat were equal for both tillage systems, while minimum-tillage spring wheat yields were 5% less than conventional tillage yields.

Each plot in the IPM experiment retained its tillage and rotation assignment for all 6 years. Furthermore, as in the UI trials, each crop in each rotation was grown in each of the 6 years. Weather over 1986 through 1991 displayed a representative mix of wet and dry years (Young et al., 1994).

The IPM main trial did not include a *continuous* no-till treatment. Two satellite IPM trials, however, provided information on the performance of continuous no-till systems. Average crop yields under continuous no-till can be compared with the mixed conservation tillage systems in the main trial (Table 9.7). Statistical inferences cannot be made since the main and satellite trials were on different

TABLE 9.7

Comparison of Average Crop Yields from Continuous No-Till Satellite Trial and Mixed Conservation Tillage Main Trial, Wheat/Barley/Pea Rotation, USDA IPM Experiment, Pullman, Washington, 1986–91

Crop, Tillage, Trial	Weed Management Level			Average	
	MIN	MOD	MAX	Yield	Index (%)
Winter Wheat		bu/ac			
No-till, Main	96	107	112	105	100
No-till, Satellite	70	—	90	80	76
Spring Barley		ton/ac			
Min-till, Main	2.2	2.5	2.6	2.4	100
No-till, Satellite	1.8	—	2.0	1.9	78
Spring Peas		cwt/ac			
Min-till, Main	17.8	20.1	20.9	19.6	100
No-till, Satellite	14.9	—	16.3	15.6	80

All yields were adjusted to typical moisture and chaff levels in the eastern Palouse: 10% for wheat, 7.5% for barley, and 5% for peas. No-till satellite yields were normalized to average weather over 1986–91. Index equals 100% for main trial; other results are proportionate.

Source: Kwon, T., Ph.D. dissertation, Washington State University, Pullman, 1993.

experimental sites. Unlike the main trials, the satellite trials included a given crop in a three-crop rotation in only 2 of the 6 years over 1986 through 1991. To adjust for the possible effects of atypical weather in the year a crop was grown, crop yields in the satellite trials were normalized to average weather over 1986 through 1991 (Kwon, 1993).

Winter wheat in the mixed tillage system had a 25 bu/acre yield advantage over winter wheat in the continuous no-till system. Minimum-tillage spring barley and spring peas had 0.54 ton and 4 cwt/acre yield advantages over these crops grown under no-till (Table 9.7).

When comparing a winter wheat/winter wheat/spring wheat rotation grown both in the main trial and in a continuous no-till satellite trial, the mixed tillage system of the main trial again produced higher crop yields. Winter wheat yields averaged about 10 bu/acre higher and spring wheat a full 20 bu/acre higher in the main trial. The first-year satellite no-till spring wheat yields were low because of poor stand establishment. Tillage was the major difference between the main trial and satellite trial, other than possible site differences. The main trial utilized no-till only 1 year out of 3 within each rotation, whereas the satellite trial consisted of continuous no-till. A complete examination of the factors underlying the yield differences between the main trial and satellite trials has not been conducted. Increased disease incidence and weed populations in the continuous no-till system may explain some of the difference.

TABLE 9.8

Composition of Average Production Costs by Tillage System for Two Crop Rotations, USDA IPM Experiment, Pullman, Washington, 1986–91, $/acre

Cost Component	Conservation Tillage		Conventional Tillage	
	WWW	WBP	WWW	WBP
Tillage	15.55	18.30	29.91	29.68
Planting	26.36	35.96	20.07	29.66
Fertilizer	47.61	30.26	48.37	30.26
Weed control	36.44	36.77	29.54	29.60
Land cost	77.34	74.83	82.06	68.89
Harvest, other expenses	62.86	78.18	56.72	79.04
Total	266.16	274.30	266.67	267.13

WWW = winter wheat/winter wheat/spring wheat rotation; WBP = winter wheat/spring barley/spring pea rotation.

Source: Kwon, T., Ph.D. dissertation, Washington State University, Pullman, 1993.

9.4.4.2 Economic and Environmental Comparisons

The multidisciplinary IPM research project included an evaluation of the relative profitability and riskiness of the farming systems examined. Both the physical productivity and the production costs of a farming system determine its profitability. Total production costs are essentially equal for the two tillage systems in the monoculture wheat rotation, and differ by only $7/acre for the wheat/barley/pea rotation (Table 9.8). The higher tillage costs for conventional tillage systems are offset by higher planting and herbicide costs for conservation tillage. Land costs are greater for higher-yielding systems in a crop-share, land-rent arrangement, which is typical in the region. Painter et al. (1992) and Kwon (1993) provide more-detailed breakdowns of conservation and conventional tillage costs.

The relative performance of complete cropping systems can be evaluated by using a detailed economic analysis. Young et al. (1994) have summarized the average profitability and year-to-year riskiness (mostly due to weather) of the 12 farming systems examined in the IPM main trial (Table 9.9, Figure 9.1). Based on the 6-year experimental results, the wheat/barley/pea conservation tillage systems under moderate or maximum weed management dominate all other systems in terms of high average profitability and low year-to-year profit variability. These systems enjoy a $7 to $55/acre profitability advantage over their competitors. These "bottom line" advantages confirm that conservation farming systems can be developed that succeed agronomically, economically, and environmentally.

A closer examination of year-to-year profitability for conservation vs. conventional systems provides insight into the reasons conservation tillage reduces economic risk (Figure 9.2). The first 3 years of the IPM trials were relatively dry, but the moisture-conserving advantages of conservation tillage protected crop yields and

TABLE 9.9
Average Profitability (net returns over total costs) by Cropping System, USDA IPM Experiment Yield Results, Pullman, Washington, 1986–91

Rotation	Tillage	WML	Profit Ranking	Average Profit ($/ac)
WW-SB-SP	Cons.	MAX	1	15
WW-SB-SP	Cons.	MOD	2	13
WW-SB-SP	Conv.	MOD	3	–6
WW-SB-SP	Conv.	MAX	4	–10
WW-SB-SP	Cons.	MIN	5	–13
WW-SB-SP	Conv.	MIN	6	–16
WW-WW-SW	Cons.	MOD	7	–20
WW-WW-SW	Conv.	MIN	8	–25
WW-WW-SW	Cons.	MAX	9	–26
WW-WW-SW	Cons.	MIN	10	–30
WW-WW-SW	Conv.	MOD	11	–34
WW-WW-SW	Conv.	MAX	12	–40

Based on projected 1991–95 average crop prices, farm programs, and production costs. WW-SB-SP = winter wheat/spring barley/spring pea rotation; WW-WW-SW = winter wheat/winter wheat/spring wheat rotation; Cons. = conservation tillage; Conv. = conventional tillage; MIN, MOD, MAX = minimum, moderate, and maximum weed management levels (WML), respectively.

Source: Young, D. L. et al., *J. Soil Water Conserv.*, 49, 581–586, 1994. With permission.

incomes during these drought years. Moisture was more abundant during 1989 through 1991. Net returns differed little between the two systems in 1989 and 1990. Moisture was adequate in 1991, also, but cold weather in late 1990 damaged many winter wheat stands. In the IPM trials, however, the winter wheat under conservation tillage withstood winter damage and sustained stands into the spring better than that under conventional tillage. This likely accounts for the economic superiority of conservation tillage in 1991.

The economic analysis presented in Table 9.9 and Figures 9.1 and 9.2 took into account the influence of prevailing government programs on the profitability of different farming systems (Young et al., 1994). The wheat/barley/pea rotation complied with the program crop base acreages of most eastern Palouse growers during 1986 through 1991, which made it eligible for program participation, while the continuous wheat rotation did not comply (Halvorson, 1991). The move to decoupled payments and initially higher wheat prices in the 1996 Farm Bill era narrowed the profit advantage of the diversified rotation, but the underlying yield advantage of this rotation will maintain its economic advantage unless wheat prices rise to and remain at very high relative levels. The actual profitability comparisons for different growers, of course, will vary depending upon the growers' particular yields, production costs, applicable farm program factors, and anticipated crop prices.

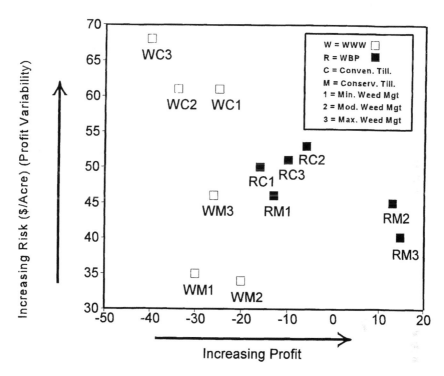

FIGURE 9.1 Comparison of Profitability and Risk for 12 Cropping Systems, USDA IPM Experiment, Pullman, Washington, 1986–91.

Source: Young, D. L. et al., *J. Soil Water Conserv.*, 49, 581–586, 1994. With permission.

The wheat/barley/pea conservation tillage system with good weed control also was environmentally sound. This system met the surface residue requirements for compliance as determined in a field evaluation by the Natural Resources Conservation Service, while the conventional tillage systems did not. Based on field measurements, chisel-plowed stubble left 47% of the crop residue on the surface compared to 9% after conventional tillage treatments (McCool et al., 1991). Conservation tillage across all rotations significantly boosted earthworm populations and microbial life in comparison with conventionally tilled plots (A. Ogg and A. Kennedy, USDA-ARS, Pullman, Washington, personal communications, 1992).

9.4.4.3 Research Comparisons

What factors underlie the relative success of the IPM conservation tillage systems experiment compared with the earlier conservation tillage systems research during the 1970s? Several characteristics are likely involved.

First, the level and effectiveness of pest management, especially weed control, improved. Second, use of improved no-till planting equipment with proper seed and fertilizer placement, and chisel plowing in the conservation tillage system, appear to have reduced soil compaction, nutrient deficiencies, the incidence of root diseases, and other stresses associated with high residue levels.

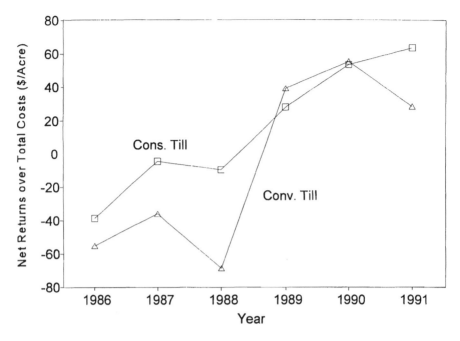

FIGURE 9.2   Profitability Comparison Between Conservation Tillage WBP and MAX WML and Conventional Tillage WBP at MOD WML, Both in Farm Programs, USDA IPM Experiment, Pullman, Washington, 1986–91.

Finally, the conservation tillage combination of minimum tillage after high-residue crops and no-till after lower-residue crops in the IPM trial utilized each tillage in situations where it provided the most benefit and least risk. A continuous no-till system for all crops in a rotation has not been commonly used, or recommended, in the region. These factors, as well as potential site differences, probably explain the contradictory findings between the UI STEEP trials and the IPM experiment.

9.5   COMPARISON WITH CURRENT FARM PRACTICES

Over 90% of the farms in the Washington–Idaho Palouse contain highly erodible land. Nearly all of these farms have developed USDA-SCS-approved conservation compliance plans, which rely heavily on increasing surface residue levels. Most growers use minimum- or reduced-tillage systems to meet residue management requirements. Many plans combine residue management with divided slope farming. No-till systems are used on a relatively small portion of the cropland. Almost all growers use herbicides as a key tool for weed control.

Conservation plans were to have been fully implemented by December 31, 1994, for growers to retain eligibility for government farm programs. However, surface residue requirements were required before that deadline. While some winter wheat acreage may not yet fully meet the requirements, most growers are making a transition to conservation farming systems. In some areas, residue requirements have been reduced by USDA to match better the performance of the best available technology.

Eastern Palouse growers are moving toward rotations with a higher frequency of small grains, which boost residue levels. In a 1989 Palouse farm survey, growers indicated equal use of a 2-year wheat/pea or lentil rotation and a winter wheat/spring barley/pea or lentil rotation (Halvorson, 1991). The 2-year rotation had dominated a decade earlier.

9.6   PROSPECTS FOR THE FUTURE

Pacific Northwest wheat growers have been slower to adopt no-till and other reduced-tillage practices than midwestern corn–soybean growers. Unfortunately, the steep topography, the occurrence of winter precipitation on frozen ground, and the dominance of winter wheat make effective conservation farming more urgent for the Palouse than for many other regions. Both on-site and off-site damages from soil erosion in this region are high and will not be permitted to continue indefinitely.

This review of the evolution of conservation farming systems research for the Palouse region concludes with good news. The combination of a diversified crop rotation (wheat/barley/pea) with no-till following peas and minimum tillage after high-residue cereal grains along with adequate weed management produced a system with superior agronomic, economic, and environmental results. While a 100% no-till system could conserve more soil, its yield performance, with current technology, remained substantially below that of a mixed conservation tillage system.

Although a successful conservation farming system has been identified for the annual cropping zone of the Washington–Idaho Palouse, substantial work remains on the part of growers, industry, and researchers in adapting and modifying such a system to meet the needs of growers in other regions. Even within the extensive annual cropping regions of the dryland Pacific Northwest, individual growers will likely need to adapt the successful IPM system to their own crop rotations and production environments.

Researchers should be encouraged to help identify systems for other cropping regions. Unfortunately, there is no substitute for long-term testing such as that exemplified by the UI STEEP trials and the USDA IPM trials in making fundamental research breakthroughs. However, on-farm testing by research groups and growers themselves can help to refine and adapt basic systems to growers' needs. The primary lesson from conservation research in the Palouse is the importance of a systems approach, which addresses all the characteristics of a successful conservation farming system. These include pest management, stand establishment and winter survival, soil fertility and moisture management, control of production costs, compatibility with government commodity programs, consistent profitability, and compliance with soil, water, and air quality targets. The successful IPM conservation farming system shows that these multiple objectives can be achieved by an appropriately diversified crop rotation, a mix of conservation tillage practices suitable to the crops in the rotation, close attention to pest management, especially weed and disease control, and an adequate fertility management strategy. The moisture-conserving and stand protection features of a well-designed conservation farming system will often stabilize and increase yields during years with adverse weather.

Removal of planting restrictions in the 1996 Farm Bill will make it easier for farmers to adopt agronomically sound conservation farming systems. Farmers should also benefit from the significant research progress over the past three decades in developing and disseminating successful conservation farming systems in the Pacific Northwest. During the next decade, even more progress is likely, but its rate will be strongly influenced by the innovativeness and cooperation among farmers, researchers, agribusiness, and policymakers.

ACKNOWLEDGMENTS

Douglas L. Young is a professor of agricultural economics at Washington State University, Pullman, WA; Frank Young is a research agronomist with USDA-ARS in Pullman, WA; John Hammel is a professor of soil science at the University of Idaho, Moscow, ID; and Roger J. Veseth is an extension conservation tillage specialist at the University of Idaho, Moscow, ID. The authors thank Elizabeth Kirby and David Granatstein for editorial assistance.

REFERENCES

Boerboom, C.M., F.L. Young, A.G. Ogg, T. Kwon, and T. Feldick, 1993. IPM Research Project for Inland Pacific Northwest Wheat Production, Washington State University Agricultural Experiment Station Research Bulletin XB 1029, Pullman.

Cochran, V.L., L.F. Elliott, and R.I. Papendick, 1981. Effect of crop residue management on yield and water-use efficiency of direct drilled winter wheat, unpublished manuscript, USDA-ARS and Agricultural Research Center, Washington State University, Pullman.

Cook, R. J. and R. J. Veseth, 1991. *Wheat Health Management*, APS Press, St. Paul, MN.

Halvorson, R.L., Jr., 1991. Palouse Agriculture: Farm Characteristics and Farmer Perceptions on Policy Alternatives, M.A. thesis, Department of Agricultural Economics, Washington State University, Pullman.

Hammel, J.E., 1989. Long term tillage and crop rotation effects on bulk density and soil impedance in northern Idaho, *Soil Sci. Soc. Am. J.*, 53, 1515–1519.

Hammel, J.E., 1993. Influence of long term tillage and rotation practices on winter wheat production, unpublished manuscript, Department of Plant, Soil, Entomological Science, University of Idaho, Moscow.

Hammel, J. E., K. E. Saxton, D. K. McCool, R. I. Papendick, J. F. Kenny, and H. Kok, 1987. Structure and compaction of soils in relation to conservation tillage systems, in L.F. Elliott, Ed., *STEEP—Conservation Concepts and Accomplishments*, Washington State University, Pullman, 109–124.

Harder, R. W., C. L. Peterson, and E. A. Dowding, 1980. Influence of tillage on soil nutrient availability, physical and chemical properties, and winter wheat yield in northern Idaho, paper presented at the Tillage Symposium, Bismarck, ND.

Koehler, F.E., V.L. Cochran, and P.E. Rasmussen, 1987. Fertilizer placement, nutrient flow, and crop response in conservation tillage, in L.F. Elliott, Ed., *STEEP—Conservation Concepts and Accomplishments*, Washington State University, Pullman, 57–66.

Kwon, T., 1993. Bioeconomic Decision Models for Weed Management in Wheat, Barley, and Peas: An Econometric Approach, Ph.D. dissertation, Department of Agricultural Economics, Washington State University, Pullman.

McCool, D. K., F. L. Young, and R. I. Papendick, 1991. Influence of crop and tillage on residue cover, in Integrated Crop Management for Cereal Legume Production in the Palouse, Tech. Report 91-3, Department of Crop and Soil Sciences, Washington State University, Pullman.

Painter, K M., D.M. Granatstein, and B.G. Miller, 1992. 1992 Alternative Crop Rotation Enterprise Budgets, Eastern Whitman County, Washington, Cooperative Extension Bulletin EB1725, Washington State University, Pullman.

Peterson, C.L., E.A. Dowding, and R.W. Harder, 1979. Chisel-Planter: An Experimental Till-Plant Erosion Control System for the Palouse, ASAE Tech. Paper 78-101, St. Joseph, MI.

Reinertsen, S.A., A.J. Ciha, and C.F. Engle, 1983a. Yields of Four Spring Barley Varieties in Conventional, Minimum and No-Till Systems, Cooperative Extension Current Information Series No. 687, University of Idaho, Moscow.

Reinertsen, S. A., A. J. Ciha, and C. F. Engle, 1983b. Yields of Four Spring Wheat Varieties in Conventional, Minimum and No-Till Systems, Cooperative Extension Current Information Series No. 689, University of Idaho, Moscow.

Young, D. L., D. L. Hoag, H. R. Hinman, and R. W. Harder, 1984. Yields and profitability of conservation tillage in the eastern Palouse, Washington State University Agricultural Experiment Station Research Bulletin XB 0941, Pullman.

Young, D.L., T.J. Kwon, and F.L. Young, 1994. Profit and risk for integrated conservation farming systems in the Palouse, *J. Soil Water Conserv.*, 49, 581–586.

10 Conservation Policy Issues

David J. Walker and Douglas L. Young

CONTENTS

10.1 EVOLUTION OF U.S. CONSERVATION AND COMMODITY POLICY

Natural resource and environmental programs have been a part of U.S. farm policy since the creation of the Soil Erosion Service within the Department of Interior in 1933. The policy of helping farmers contend with the vagaries of weather, yield and prices through price and income support began nearly a decade earlier. This income support policy received more attention than environmental policy until the 1985 and subsequent Farm Bills. In the 1980s, it became clear to many that the farm commodity programs were working at cross purposes with soil conservation and water quality objectives and that changes were needed. This chapter reviews the evolution of U.S. farm commodity and environmental policy and outlines the conflict between commodity programs and environmental objectives. Research on these issues completed under the Pacific Northwest Solutions to Economic and Environmental Problems (STEEP) project is highlighted.

Government policy to combat soil erosion began in 1933 when the Soil Erosion Service, using public works funds and Civilian Conservation Corps (CCC) workers, developed demonstration projects and established cooperative agreements with farmers to provide technical assistance and labor and materials to implement conservation plans on the farms. In 1935, Congress passed the Soil Conservation Act, and President Roosevelt transferred the Soil Erosion Service to the USDA where it became the Soil Conservation Service (SCS). Cost share incentives to encourage farmers to adopt conservation practices began with the establishment of the Agricultural Conservation Program in 1936.

The SCS, with the responsibility for organizing and conducting a national soil and water program, has provided technical assistance to landowners and farm operators through locally organized conservation districts. In 1990, there were about 3000 districts covering more than 2 billion acres of farmland. The SCS was renamed the Natural Resource Conservation Service in 1995.

With passage of the Federal Water Pollution Control Act in 1972, concern about water quality intensified the importance of controlling erosion and sediment. Section 208 of the 1972 Water Quality Amendments was the first effort to control nonpoint-source pollution such as runoff from agricultural land. Congressional resolve to improve the control of nonpoint sources was reiterated in the 1977 Clean Water Act, soil conservation compliance provisions of the 1985, 1990, and 1996 Farm Bills, and the Water Quality Act of 1987. These acts provide economic incentives for the implementation of best management practices to reduce nonpoint residuals from agricultural areas. The importance of controlling erosion had shifted from productivity impacts exclusively to include sediment and water quality impacts. The Food Security Act, or 1985 Farm Bill, was landmark legislation that linked income support and environmental policy by establishing conservation compliance for farmers to be eligible for support payments. This new direction was continued in subsequent farm bills. These will be discussed in more detail later in the chapter after describing the problem they address.

The roots of farm price and income support policy go back to 1924 and the McNary-Haugen plan. This plan, motivated by the farm depression of the 1920s, attempted to raise farm prices to the "parity" level by limiting domestic supplies and dumping the surplus on foreign markets. With the stock market crash of 1929 and the ensuing worldwide economic depression, farm prices fell 50% in 3 years. The New Deal response was the Agricultural Adjustment Act (AAA) of May 12, 1933. Title I of the AAA provided for price and income supports with acreage controls to reduce supply.

In this early period, farm policy stressed price supports. The government guaranteed a market price above the market clearing level and maintained it by purchasing, in effect, the excess supply that occurred at the supported price. To prevent oversupply, spurred by high prices, the government attempted to limit production. One early method of supply control that has relevance for current policy was the "soil bank" program enacted into law with the Agricultural Act of 1956. The first part of this Act was the Acreage Reduction Program (ARP) that retired a specified acreage from production each year. This land that was put in the soil bank for 1 to 3 years was not eligible for harvesting a crop or for pasturing. The second part of the Act established a Conservation Reserve Program (CRP) that was designed to shift low productivity land into long-range conservation uses or the soil bank. Both programs were relatively short-lived, the first ended in 1958 and the second was not promoted after 1959. Both programs would re-emerge in the 1980s and will be discussed in more detail later in this chapter.

Most attempts to limit production were calculated from historical levels of crop production or "base acres." The concept of base acres would prove to be an impediment to soil and water conservation in the decades ahead. Another inconsistency was for an agricultural exporting country to support prices above world equilibrium levels.

With the Food and Agricultural Act of 1965, the concept of price supports was abandoned in favor of market-clearing prices with farm income supported through government deficiency payments. Another aspect of this transformation, the Agricultural and Consumer Protection Act of 1973, set the stage for subsequent farm income support policy. This Act established the concept of target price as a basis for calculating deficiency payments. The government allowed prices for program crops to seek their market-clearing levels but made deficiency payments to farmers equal to the difference between the legislated target price and the market price or the loan rate, whichever was higher. The major commodity program crops were food grains, feed grains, and cotton. Again, to limit the government financial liability, farmers were allowed payment only if they restricted planting of commodity crops to historical base acreage less any set-aside acres. Farmers were required to set-aside a specified percentage of the base acreage each year depending on supply and demand projections. If supplies were tight, farmers were allowed to plant all of their base acres in commodity crops. If carryover stocks were high, set-aside was increased, in some years to nearly 30%.

This method of supporting farm income placed great importance on historical acreage in program crops or base acres. This structure thwarted conservation efforts. Some farmers sought to increase base acres in commodity program crops by planting on marginal land that might be highly erodible. The expanded base acreage increased entitlements to future deficiency payments and also allowed farmers subsequently to comply with set-aside by using land with a lower opportunity cost of lost production. The importance of maintaining base acres in wheat or other program crops left little room for soil-conserving crops in a rotation (Beus et al.; 1990). The planting of marginal land to program crops and the reluctance to use soil conserving crop rotations contributed to high soil erosion rates. Using yield boosting summer fallow with inadequate conservation measures on set-aside acreage also increased soil erosion, especially in higher precipitation regions which normally would be annually cropped. In conclusion, commodity programs and soil conservation programs were operating at cross purposes; conservation programs attempted to reduce erosion while commodity programs unintentionally increased erosion.

The Food Security Act of 1985 (FSA) attempted to deal with this dissonance. This Act established conservation compliance, the Conservation Reserve Program (CRP), and the Sodbuster and Swampbuster programs. Conservation compliance requires farmers with highly erodible land to conserve soil by linking eligibility for commodity program payments to soil conservation. To be eligible, farmers had to develop approved conservation plans by 1990 for highly erodible land. For continued eligibility for all USDA farm programs, these plans were required to be fully implemented by 1995.

In return for an annual rental payment from the government, the CRP removes highly erodible land from production for a contract period of 10 years. The program was designed to reduce erosion and sedimentation, to curb production surpluses and to provide income support to growers. Farmers have to plant a conservation cover crop like grass on their CRP acres. The CRP and conservation compliance programs were continued in the 1990 and 1996 Farm Bills.

The sodbuster provision requires that if previously unplanted highly erodible ground is planted, a conservation system must be applied or program eligibility is lost. The swampbuster provision, in general, prevents farmers from draining and planting naturally occurring wetlands. This provision hopes to stem the decrease in flood control, water quality, fish and wildlife habitat, and recreational opportunities that occur when wetlands are drained. The 1985 Farm Bill (FSA) also froze program yields on which deficiency payments are made. This limits the profitability of program crops which may have high erosion and dilutes the incentive for heavy fertilizer applications to boost yields. This bill also restricted base acreage growth.

The 1990 Farm Bill ameliorated base acreage disincentives to soil-conserving crop rotations by allowing farmers the flexibility to plant a portion of their base acres to other crops without losing base acreage. This bill further restricted base acreage growth and emphasized targeting of CRP to lands with multiple environmental benefits, such as both soil and water quality improvement. The 1996 Farm Bill strengthened incentives to target CRP enrollment to environmentally fragile lands. The 1996 CRP legislation ranked CRP bids in descending order based on the ratio of expected environmental benefits to government costs. Environmental benefits were broadened to include wildlife habitat protection, surface and groundwater quality protection, and wind erosion reductions as well as prevention of water-based erosion.

With regard to commodity programs, the 1996 Farm bill decoupled farm payments from cropping decisions and market prices. Farmers receive contractual transition payments related to historical acreage and yields (both of which are frozen) and these payments will be phased out in 2002. Essentially there is 100 percent cropping flexibility, as farmers can plant any crops except fruits and vegetables and be eligible for these payments. Base acreage disincentives for conservation are totally eliminated and market price incentives in cropping decisions are restored.

10.2 STEEP RESEARCH ON COMMODITY POLICY IMPACTS ON CONSERVATION

A regional study with national policy implications directly examined the issue of base erosion with soil- and water-conserving rotations. A study by Young and Painter (1990) examined the criticism that the 1985 FSA encouraged inflexible cropping patterns dominated by program crops. Because grain crops in the commodity program have high erosion potential and most of the major program crops tend to be intensive users of agrichemicals, farm programs have also been blamed for discouraging low-input sustainable agricultural practices (National Research Council, 1989; Ek, 1989).

Growers may be deterred from including in rotations hay, edible legumes, green manure crops, or other crops that reduce erosion and/or the use of fertilizer and pesticides because these crops reduce the base acreages of program crops. A major shortcoming of farm programs, including the 1985 FSA, was the emphasis on support of specific commodity program crops, discouraging cropping flexibility necessary for soil- and water-conserving rotations. The 1990 Farm Bill allowed farmers the flexibility to plant 15 to 25 percent of their base acreage to specified other crops without losing base acreage for future program payments. However, deficiency payments were not made on the normal (15% of base acres) flexible acres. as discussed above, the 1996 Farm Bill removed most planting restrictions except for quota crops like tobacco and peanuts.

In the Bush Administration's Normal Crop Acreage (NCA) plan, which was rejected in the 1990 Farm Bill, the entire base acreage would have been flexible and eligible for deficiency payments. Young and Painter examined the impact of the 1985 FSA and the proposed 1990 NCA plan on the relative profitability of a conventional and green manure rotation in the southeastern Washington study area. Specifically, the study examines the acreage reduction effect (ARP), deficiency payment effect (DP), and base effect. The base effect measures the present value of changed (lost) future deficiency payments from changed (reduced) base acreage in the future due to present (reduced) planting decisions. The DP effect per rotational acre equals the deficiency payment per acre for each crop times the proportion of each crop in the rotation. The ARP effect equals the weighted sum across program crops of the direct cost of maintaining each crop's set-aside plus the opportunity cost of forgone net returns from crop production on set-aside. The weights correspond to the proportion of rotational acreage in set-aside for each crop.

The rotations compared under the two policies were a conventional 4-year grain intensive rotation, winter wheat/spring barley/winter wheat/dry peas, and a low-input 3-year rotation, dry peas/ winter wheat/ unharvested medic green manure. Medic, a biennial legume, offers nitrogen fixing advantages and appears to reduce wheat root diseases. In the low-input rotation known as PALS (Perpetuating Alternative Legume System), wheat which follows 2 years of legumes is grown without commercial fertilizers, herbicides or fungicides (Goldstein and Young, 1987). Peas, however, received normal pesticide applications.

The FSA was analyzed for 5 years, 1986 through 1990. In 3 of the 5 years the FSA decreased PALS's relative profitability compared to the conventional rotation. In 1986, 1987, and 1990 the ratio of PALS profitability to the conventional rotation fell under the FSA. In the other 2 years the FSA enhanced PALS' relative profitability (see Table 10.1). The profitability disadvantage for PALS in three years was due to a combination of generous deficiency payments and modest ARP opportunity costs for the competing grain-intensive rotation. Opportunity costs decrease from low market prices of crops retired on set-aside acres or low set-aside rates. In contrast, larger ARP opportunity costs and modest deficiency payments explain the enhancement of relative profitability for PALS under the FSA in the other 2 years. In general, the relative profitability of green manure rotations is hurt by high deficiency payments and low ARP costs, both direct and indirect.

Turning to the base effect, long-term use of the conventional rotation will generate a barley base and wheat base of 25% and 50% of farm acreage, respectively. These bases, upon switching to PALS, will decrease because the entering base is computed each year as the average acreage planted and considered planted (in ARP) over the previous 5 years. Because the PALS rotation contains no barley, the barley

TABLE 10.1
Annual Net Returns, ARP Effects, and DP Effects
($/acre of rotation) for PALS and Conventional Rotations
Due to 1985 FSA

Year	Rotation and Ratio[a]	Net Returns over Variable Costs		DP Effect	ARP Effect	Net Effect[b]
		No FSA	With FSA			
1986	PALS	64	133	68	0	68
	CONV	51	144	103	–10	93
	Ratio	*1.25*	*0.92*			
1987	PALS	64	119	56	0	56
	CONV	59	119	75	–16	59
	Ratio	*1.08*	*1.00*			
1988	PALS	103	122	18	0	18
	CONV	124	113	20	–32	–12
	Ratio	*0.83*	*1.08*			
1989	PALS	94	102	8	0	8
	CONV	107	108	12	–11	1
	Ratio	*0.88*	*0.94*			
1990	PALS	73	110	37	0	37
	CONV	57	116	62	–2	60
	Ratio	*1.28*	*0.95*			
1986–1990	PALS	80	117	37	0	37
	CONV	80	120	54	–14	40
Mean	*Ratio*	*1.00*	*0.98*			

[a] Ratio of net returns of PALS to the net returns of conventional (CONV) for the year and program scenario.
[b] Because of rounding errors, net effect computed as the sum of the DP and ARP effect may differ slightly from the same effect computed as the difference between with FSA returns and no FSA returns.

Source: Young and Painter, 1990.

base disappears after 5 years. The wheat base under the FSA declines from 50 to 41% in 5 years (Table 10.2). It is not possible to calculate the exact present value of the base loss because future setaside rates, target prices and market prices are not known. This loss due to the base effect reduces the long-term profitability of green manure rotations.

The NCA proposal would have allowed base acreage flexibility for the entire farm and would have paid deficiency payments on flexible acres. The expanded base flexibility prevents the base erosion problem inherent in the FSA. Under the FSA, total wheat and barley base eroded from 75 to 41% of farm acreage after 5 years of PALS, while NCA preserved base at 67% of farm acreage.

Comparing the relative profitability of PALS and the conventional rotation under the NCA proposal, PALS was more profitable in 4 of the 5 years (see Table 10.3).

TABLE 10.2
Wheat and Barley Annual History and Base Acreage for
PALS Rotation under Two Policy Scenarios

Policy and Crop	Proportion of Total Acreage, for Year:					
	1986	1987	1988	1989	1990	1991
FSA						
Wheat						
Entering base	0.50	0.49	0.48	0.47	0.44	0.41
Annual history[a]	0.43	0.46	0.46	0.37	0.35	—
Barley						
Entering base	0.25	0.20	0.15	0.10	0.05	0
Annual history[a]	0	0	0	0	0	—
NCA Proposal						
Program crops plus permitted uses:						
Entering base	0.75	0.73	0.72	0.70	0.69	0.67
Annual history[a]	0.67	0.67	0.67	0.67	0.67	—

[a] Planted and considered planted in current year.

Source: Young and Painter, 1990.

By contrast, under FSA, PALS was more profitable in only 2 years. The NCA proposal with base flexibility benefits resource conserving rotations like PALS. The results of this study supported expanding base flexibility in future farm programs. As mentioned above, this subsequently happened in the 1990 and 1996 farm Bills. In 1990, 15 to 25 % of base acreage was made flexible and in 1996 the entire concept of base acreage was abandoned.

10.3   STEEP RESEARCH ON 1985 FOOD SECURITY ACT CONSERVATION PROVISIONS

Another regional analysis of the 1985 FSA with national implications was conducted by Young et al. (1991). This study looked at the equity and efficiency aspects of the Conservation Compliance and Conservation Reserve Programs in the Palouse region. Equity within these FSA programs involves two issues: (1) the distribution of conservation costs between farmers and taxpayers, and (2) the distribution of costs among farmers with different land endowments. Efficiency refers to the cost effectiveness of the program, the total cost per ton of soil conserved. Soil conserved was used as a proxy for total on-site and off-site damage averted.

The study examined the impact on farmers and taxpayers of alternative conservation compliance levels and CRP bid caps for Whitman County, Washington. A farm-level, mixed- integer programming model was used to simulate farmer profit-maximizing responses to alternative provisions of the 1985 FSA. This model was

TABLE 10.3
Net Returns and Difference in Deficiency Payments
($/acre of rotation) for Two Policies by Rotation

Year	Rotation and Ratio[a]	Net Returns FSA	Net Returns NCA	Difference in DP (NCA–FSA)
1986	PALS	133	167	34
	CONV	144	144	0
	Ratio	0.92	1.16	
1987	PALS	119	138	19
	CONV	119	119	0
	Ratio	1.00	1.16	
1988	PALS	122	123	1
	CONV	113	113	0
	Ratio	1.08	1.09	
1989	PALS	102	105	3
	CONV	108	108	0
	Ratio	0.94	0.97	
1990	PALS	110	134	24
	CONV	116	116	0
	Ratio	0.95	1.16	
1986–1990	PALS	117	133	16
Mean	CONV	120	120	0
	Ratio	0.98	1.11	

[a] Ratio of net returns, PALS/CONV, for the year and program scenario.

Source: Young and Painter, 1990.

applied to 3 representative farms which reflect farm size, land characteristics, and crop yields of the western, central, and eastern subregions of the Palouse region. Annual precipitation varies across subregions with 12 to 15 in. in the western, 15 to 18 in. in the central, and 18 to 22 in. in the eastern subregion.

The distribution of farmer and taxpayer costs of the Conservation Reserve Program and conservation compliance were estimated by comparing farmers' profits and taxpayers' costs for the benchmark solution of the profit-maximizing models without conservation provisions versus the solutions with the various conservation compliance standards and CRP rents. Farmer costs for each policy mix was the difference in farm income between the given policy and the benchmark. Government costs were the difference in government outlays between the given policy and the benchmark.

Although the data and assumptions in this study are several years old, the results of this study illustrate the effects of incentives designed into the 1985 FSA CRP and compliance provisions. Results in Table 10.4 present the distribution of costs between farmers and taxpayers for the three subregions. In the eastern subregion, with a 1T soil loss standard and $60 per acre CRP rent, compliance was projected to cut erosion by 60%. Over half of the annual cost was borne by farmers who incurred $1.90/ton of soil saved versus $1.44/ton for taxpayers. In the lower rainfall western and central subregions, the combination of CRP rents and compliance standards generally increased the projected farmer's net return due to a net (CRP) subsidy from the government. Only at the strict 1T compliance level in the central subregion, were farmers estimated to bear any cost of conservation. This result occurs because of the low opportunity cost and thus high projected profitability of CRP in these less productive subregions.

Table 10.4 also shows the impact of varying CRP rates assuming a constant 1 to 2 T compliance standard. In all subregions, CRP was a profitable way for farmers to meet conservation compliance at $40, $60 and $80/acre CRP rents. Therefore, taxpayers incur some of the cost of soil conservation. Information on the distribution of costs among farms is also included in Table 10.4. With a uniform bid cap of $60/acre, farmers in the eastern subregion bear 64% of conservation costs. In the central and western subregions farmers receive a net subsidy from the conservation programs. The projected gains from CRP payments in less productive subregions illustrate the potential for improving cost effectiveness by differentiating bid caps based on land productivity and opportunity cost.

This study draws two important policy conclusions. The uniform multicounty bid caps contributed to the inequity and low cost effectiveness of erosion control under the 1985 FSA. Projected net costs to farmers in the three subregions under the program ranged from an $8/acre reduction in net income in the east to a $9/acre gain in the west. Cost effectiveness to taxpayers dropped threefold from the eastern to the western subregion. One solution to the uniform bid cap problem would be a true competitive bid system for CRP enrollment (U.S. GAO, 1989). Young, Walker, and Kanjo also showed that tightening compliance standards was a more cost-effective means of reducing erosion than increasing CRP rents. This underscores the concern over relaxing conservation compliance as happened during the 1985 Farm Bill era with the introduction of alternative conservation plans in place of soil loss tolerance criteria.

A later study compared the cost effectiveness of the more environmentally targeted 1990 CRP to the 1985 CRP in the Washington, Oregon, Idaho, and California (Young, Bechtel, and Coupal, 1994). The 1990 CRP was shown to be more effective in concentrating CRP enrollment in counties with more erodible cropland. The fixed multicounty bid caps in the 1985 CRP appeared to direct enrollment to counties with low agricultural productivity and low to modest erodibility. However, due to program and funding constraints, the 1985 CRP enrolled 34 million acres compared to only 2.5 million acres for the 1990 CRP.

10.4 STEEP LAND RETIREMENT POLICY RESEARCH

A study by Ellis et al. (1989) examined a local SCS watershed program for encouraging participation in an erodible land "grass-out" program (Grassout). Where crop productivity and opportunity cost of participation are high, regular CRP enrollment tended to lag. In Whitman County, Washington, less than 7% of the target CRP sign-up set by ASCS had been enrolled by late 1988. The local conservation district,

TABLE 10.4
Impact of Conservation Compliance Standards and CRP Rates on Soil Conserved and Distribution of Costs

Palouse Subregion	Conservation Compliance Standard	Gross CRP Rate ($/ac)	Soil Conserved (ton/ac)[a]	Cost Changes ($/ac)				Cost of Soil Conserved ($/ton) Borne by	
				Taxpayers		Total	Farmers[b]	Farmers	Taxpayers
				Deficiency Payments	CRP				
East	1T	60	6.8	-13.32	23.16	9.84	12.92	1.90	1.44
	1-2T	60	5.3	-6.02	10.49	4.47	7.95	1.50	0.84
	2T	60	3.6	0	0	0	3.24	0.90	0
	None	60	0	0	0	0	0	c	c
	1-2T	None	4.7	-4.86	0	-4.86	20.54	4.37	-1.01
	1-2T	40	5.3	-6.02	6.99	0.97	11.45	2.16	0.18
	1-2T	60	5.3	-6.02	10.49	4.47	7.95	1.50	0.84
	1-2T	80	5.3	-14.18	32.86	18.68	3.55	0.67	3.60
Central	1T	60	5.0	-16.60	25.25	8.65	1.25	0.25	1.72
	1-2T	60	4.0	-9.83	16.65	6.82	-2.32	-0.58	1.70
	2T	60	1.6	-4.08	11.33	7.25	-7.57	-4.73	4.39
	None	60	1.5	-3.05	10.08	7.03	-8.68	-5.79	4.84
	1-2T	None	3.3	-7.88	0	-7.88	15.64	4.74	-2.38
	1-2T	40	4.0	-9.83	11.10	1.27	3.24	0.81	0.32
	1-2T	60	4.0	-9.83	16.65	6.82	-2.32	-0.58	1.70
	1-2T	80	3.9	-15.65	33.92	18.27	-9.44	-2.42	4.68
West	1T	60	3.7	-8.99	25.00	16.11	-3.37	-0.91	1.84
	1-2T	60	5.0	-10.92	25.00	14.08	-9.05	-1.81	2.79
	2T	60	4.5	-8.89	25.00	16.11	-14.22	-3.16	3.56
	None	60	4.5	-8.89	25.00	16.11	-14.22	-3.16	3.56
	1-2T	None	7.5	0	0	0	10.12	1.35	0

TABLE 10.4
Impact of Conservation Compliance Standards and CRP Rates on Soil Conserved and Distribution of Costs (continued)

Palouse Subregion	Conservation Compliance Standard	Gross CRP Rate ($/ac)	Soil Conserved (ton/ac)[a]	Cost Changes ($/ac)					Cost of Soil Conserved ($/ton) Borne by	
				Taxpayers			Farmers[b]		Farmers	Taxpayers
				Deficiency Payments	CRP	Total				
West	1–2T	40	5.7	−11.78	20.62	8.84	−5.19		−0.91	1.55
	1–2T	60	5.0	−10.92	25.00	14.08	−9.05		−1.81	2.79
	1–2T	80	8.4	−5.85	25.00	19.15	−11.76		−1.40	2.28

[a] Soil conserved equals the difference between erosion at no-conservation-policy benchmark and with the policy.
[b] Farmers' increased (decreased) cost equals reduction (expansion) in net income compared to benchmark.
[c] Undefined because changes in taxpayer and producer cost and soil conserved with the policy equal zero.

Source: Young, Walker, and Kanjo, 1991.

working with the SCS, provided incentives for a supplemental program to grass out the most highly erodible lands. This local program was evaluated in the study in conjunction with CRP. The Grassout option included benefits not offered by CRP. Grassout acreage qualified as set-aside in the farm commodity program, but CRP did not. In addition, the government paid 65% vs. 50% of establishment cost. Finally, base acreage was not reduced by participation in the Grassout Program. A drawback was that the Grassout Program terminates in 5 years. At that time the grower may leave the land in grass, reintroduce crops subject to sodbuster provisions that require reducing soil loss to tolerance levels, or possibly enroll in CRP.

Computerized farm-planning models (mathematical programming) were used to determine the minimum acceptable payment rates for Grassout under three policies covering two five-year periods. The three policies were

1. NOCRP2—CRP signup is not possible in period 2 at the end of Grassout;
2. YESCRP2—CRP signup is available in period 2 and crop base is reduced;
3. NOBASRED2—CRP is available in period 2 and base reduction is not required.

Both 10% set-aside and 20% set-aside rates were evaluated for each policy for a total of six policy scenarios. Because of the dichotomous nature of the decision to participate or not in government programs and the nonlinear nature of base reduction, a mixed-integer, separable programming model was used with a 10-year decision framework (two 5-year periods).

In the low (12-15 in.) precipitation zone, an annual payment of $153/acre for 3 years was projected to grassout the most erodible land if there is no CRP in period 2. If CRP enrollment is available at the end of the 5-year Grassout contract, the estimated payment decreased to $86/acre because the CRP payment exceeds the expected return from cropping. Increasing the set-aside percentage to 20% for barley and wheat affected only the YESCRP2 scenario where the payment increased from $86 to $90/acre.

In the intermediate-precipitation zone for all scenarios, both with and without CRP available in period 2, higher payment rates were required than in the low-precipitation zone because the opportunity cost of Grassout is higher.

10.5 STEEP WATER QUALITY POLICY RESEARCH

Taylor et al. (1992) compared the cost-effectiveness of five policies for controlling nonpoint-source pollution from nutrients using least-cost frontiers for five farms in the Willamette Valley of Oregon. The representative farms included two for river bottomland, two for broad terrace lands, and one for the foothills. The study combined a biophysical process model with an economic optimization model of farm management behavior. The costs of the various policies when simulated on the economic optimization model were compared with the least-cost solution for the appropriate level of control from the least-cost frontier generated with the economic optimization model.

The five policy options evaluated were

1. Specific taxes on leached nitrates, surface runoff of organic nitrogen and runoff of nitrates;
2. A tax on nitrogen fertilizer input;
3. A per-acre effluent standard, presumably enforced with penalties;
4. A similar requirement to use no-till drills;
5. A ban on fall fertilizer application.

The effluent taxes come closest to the least-cost solutions. However, on poorly drained soils, the tax would need to be large because alternative farming practices are not very profitable. The costs to growers of per-acre standards are higher than charges because of the restriction to reduce leachate to a uniform standard on all acres even though it may be more efficient to reduce leachate more on some acres than on others. Input taxes equal to 50 and 100% of the price of nitrogen fertilizer reduce applications only slightly because of the insensitivity of nitrogen use to price changes. The no-till directive reduces sediment but increases nitrogen leachate. A fall fertilizer ban increased leachates on most farms because production shifted away from fall-planted crops to higher-pollution spring crops. Successful pollution control policies must recognize site-specific characteristics. A single policy will not work across all farm types because the effectiveness of control policies and practices can vary.

A study by Wu, Walker and Brusven (1997) evaluated the interaction of planting flexibility and conservation compliance provisions from the 1985 and 1990 Farm Bills. Results showed that conservation compliance by itself enhanced environmental goals by reducing erosion and sedimentation. Conservation standards that are too strict can be self-defeating because growers will leave the farm program and participate in neither support payments or conservation compliance. Unconstrained by conservation compliance, the study showed growers may increase erosion. To encourage participation in farm income support and conservation compliance programs, the conservation standards must have a reasonable cost for achievement compared to the farm income support incentives offered. In this region, planting flexibility alone was projected to reduce erosion and sediment damage, but only slightly. In combination with conservation compliance, planting flexibility increased erosion and environmental damage, compared with conservation compliance alone. Thus, in this region, conservation compliance appears to be the stronger policy instrument and it is more effective when not combined with flex policy.

10.6 STEEP RESEARCH ON PROPOSALS FOR POLICY REFORM

A final study compared the projected social welfare impacts of six contemporary policy scenarios across two regions. This study, by Painter and Young (1993), compared: the **1990 Farm Bill** which provides 15% unpaid flex base; the 1990 Farm Bill with the unpaid flex base increased to **40% Unpaid Flex**; the 1990 Administration's **NCA**; **Decoupling** payments from specific annual crop production and substituting a lump sum payment based on historical levels; **Recoupling** payments to soil conservation and water quality protection; and **No Programs**. The flex base permits farmers to plant any combination of program crops and specified nonprogram crops (usually soil conserving crops) on base acres without losing base acreage. If deficiency payments are not paid for nonprogram crops on base acreage, the policy is called unpaid flex base. The **NCA** proposal allowed 100% flex base and provided deficiency payments on the total base acreage less set-aside.

The environmental issues related to agriculture differed over the two case study regions: the Washington–Idaho Palouse and the North Carolina coastal plain. In the dryland, grain-growing Palouse, soil erosion is the major environmental problem because of the steep hills. In the Palouse, 88% of the land is classified as highly erodible compared with 2% in the North Carolina coastal plain. In the corn-, soybean-, and tobacco-producing North Carolina region, nitrogen and pesticide leaching to groundwater is the major concern. Relatively high water tables, higher rainfall, sandier soils, and greater rural population density account for the greater damage from polluted groundwater. Mixed-integer programming was used to solve for the profit-maximizing crop rotations and practices for each policy in the two regions. The projected economic and environmental impacts for each farm policy are estimated as the difference from results projected under the **1990 Farm Bill** benchmark and are presented in Table 10.5.

In the Palouse, the various policies projected little overall change in social welfare as measured by total economic and environmental returns (column 7 in Table 10.5). Trade-offs among farm managers, landowners, and taxpayers occurred under each policy. Returns to management were highest under the **1990 NCA** and **recoupling**. Under the other policies, returns to management decreased with the sharpest decline occurring, as expected, for **No Programs**. Returns to land based on crop share rents decreased for all new policies because wheat production declines relative to levels projected under the status quo **1990 Farm Bill**. Projected change in "consumer surplus", or impact on consumers due to food price changes, were negligible due to the small survey-predicted crop price changes by policy. Taxpayer cost fell sharply for **No Programs** and declined more modestly for **40% Unpaid Flex**. Projected tax cost increased for **Recoupling** and the **1990 NCA** proposal. Off-site erosion damage decreased for all policies but the decline is greater for **Decoupling**, **Recoupling** and **No Programs**. On-site erosion damage also declined for all policies, with the greatest decrease for the same three policies.

In the Palouse, if transaction costs of government programs were included at estimated levels of 25-50% of deficiency payments (Alston and Hurd), **No Programs** would be superior to all policies. Taxpayers would benefit at the expense of farm managers, however.

In North Carolina, policy reform offers substantial projected gains in social welfare. In order, by decreasing magnitude, **Recoupling**, **1990 NCA**, **Decoupling**, **No Programs** and to a lesser extent the 1990 Farm Bill with **40% Unpaid Flex** provided higher social welfare than the **1990 Farm Bill**. Returns to management increased for **Recoupling**, **1990 NCA**, and **Decoupling** but decreased substantially for **No Programs** and modestly for 1990 Farm Bill with **40% Unpaid Flex**.

TABLE 10.5
Economic and Environmental Indicators When Returns to Management are Optimized, Expressed Relative to the 1990 Farm Bill, by Policy and Technology Availability, Washington–Idaho Palouse ($/ac/yr)

Policy/Rotations Available	Returns to Management (1)	Returns to Land (2)	Change In Food Price Impact (3)	Taxpayer Cost (4)	Off-Site Erosion Damage (5)	On-Site Erosion Damage (6)	Total Econ. and Env. Results (1+2-3-4-5-6) (7)
1. 1990 Farm Bill, 40% Flex							
a. All rotations	-4	-7	0.06	-8	-5	-0.38	2
b. No alt. rotations	-2	-3	0.25	-7	0	0.00	2
2. 1990 Admin. Proposal							
a. All rotations	5	-1	0.17	5	-2	-0.09	1
b. No alt. rotations	7	-4	0.41	5	1	0.11	-4
3. Decoupling							
a. All rotations	-1	-10	-0.38	0	-11	-0.84	1
b. No alt. rotations	5	0	-0.61	0	1	0.05	5
4. Recoupling							
a. All rotations	9	-10	-0.08	13	-11	-0.84	-2
b. No alt. rotations	1	0	0.03	0	1	0.05	0
5. No Programs							
a. All rotations	-15	-17	-0.23	-23	-11	-0.84	3
b. No alt. rotations	-13	-8	-0.29	-23	1	0.05	1

Source: Painter and Young, 1993.

Projected returns to land were unchanged because cash rents predominate in the region and were not affected in the short run by the polices. The change in consumer surplus was negligible because of the relatively small price impacts with the policies. Taxpayer cost decreased for **No Programs** and 1990 Farm bill with **40% Unpaid Flex**, but increased for **Recoupling** and **1990 NCA**. Environmental damage from leaching decreased with all policies.

In North Carolina, policies that offer increased planting flexibility, decoupled payments, or payments linked to environmental goals perform well because they allow profitable soybean production. Farmers no longer plant relatively unprofitable corn and wheat to receive government payments. Since these cereals are heavy nitrogen users, reduced plantings under revised polices lessen the environmental damage from nitrogen leachate.

Painter and Young suggested some important policy implications from these results. Policy reform in the North Carolina region offers both potential social welfare gains and environmental improvement. In the Palouse, the lack of alternative crops could limit the scope for welfare gains with policy reform. These divergent impacts underscore the importance of regional evaluations of national policy before undertaking policy reform. Further, policy reform alone in regions with limited cropping alternatives may not be sufficient. Development of environmentally sound cropping alternatives or sustainable farming practices would also be required to realize environmental gains under policy reform. To avoid losing environmental benefits during periods of excess demand and high grain prices when farmers would be inclined to expand plantings of erosive grains, income support payments based on environmental objectives—rather than program crop acreages—should be implemented.

10.7 CONCLUSIONS

The preceding review of recent conservation policy research in the STEEP project permits several important conclusions when combined with observation of the current agricultural scene. First, conservation policy and implementation made major strides in the past decade. The 1985 FSA conservation provisions were the most significant since the SCS (now NRCS) was created 50 years earlier. Conservation compliance is the first major legislation to make soil conservation performance an eligibility requirement for receipt of commodity program or other USDA payments.

Equally important, the FSA marked the beginning of gradual dismantling of important barriers to sound conservation practices. Program yields were frozen and restrictions on base acreage growth were included in both the 1985 and 1990 legislation. This weakened incentives to build yields and base in one or two program crops at the expense of more diversified and more resource-conserving rotations. Permitting 15 to 25% flexible base acreage in 1990 further opened the door for diversification. Flexibility was extended with the 1996 FAIR which decoupled payments from current plantings of program crops. Essentially there is complete cropping flexibility with the 1996 Farm Bill except that payments are not allowed for land planted to fruits and vegetables. Although it imposes short-run adjustment costs, the freezing of target prices in the 1990 FSA and decoupling in the 1996 Farm Bill could lead in the long run to less dependence on one or two program crops. Early

STEEP research projected that decoupling could be a cost-effective conservation policy but could be costly for farmers in periods of weak market prices.

On the whole, these policy shifts will have greater impacts in the Midwest or in Pacific Northwest irrigated regions where more cropping alternatives exist. In many dryland farming regions of the inland Pacific Northwest, there are few alternatives to wheat and barley. Even in these regions, in the long run, these policy changes might encourage development of less- resource-intensive production practices and of alternative rotations. However, the economic losses projected by early STEEP research of a **No Programs** policy for Palouse wheat and barley farmers appears to be occurring by the time of this writing (mid-1998), as the fixed 1996 Farm Bill payments are inadequate to offset exceptionally low wheat and barley market prices.

Despite recent increases in the flexibility of commodity programs, it is clear that commodity program reform alone will not solve all soil and water degradation problems. For example, many fruit, vegetable, and forage crops which fall outside the government program umbrella receive high applications of fertilizer and pesticides. These nonprogram crops also are posing threats to ground water in some areas of western Washington and Oregon and the irrigated valleys of Idaho. Conservation compliance-type legislation, which receives its power from the threat of denial of commodity program benefits, will be ineffective in regions growing non-commodity program crops. Increasing "flexibility" in commodity programs for feed and food grains has little effect on fruit and vegetable growers. The same arguments apply to livestock feeding operations which might threaten surface water quality. Local, state, and federal programs outside the standard conservation-commodity policy reform process are required to handle such situations.

Another important conclusion from the recent wave of conservation policy reform is that much can be done by improving the design and implementation of existing policies. A good example is the improvement in CRP policy in the 1990 and 1996 Farm Bills. While the 1985 CRP was very popular, it was also a very expensive program. Much of the enrollment and cost occurred in areas of marginal erodibility and relatively low productivity. Highly erodible areas in productive cropping regions were often by-passed. Examples include the highly productive and highly erodible eastern Palouse in Whitman County, Washington and Latah County, Idaho, where acreage enrollment was low. The 1990 and 1996 extensions of the CRP targeted the program to regions of high erosion or water quality vulnerability. The new CRP also permitted program managers greater flexibility in per-acre rental payments. These changes have increased enrollment in priority areas, although the total number of acres enrolled and program cost substantially declined. Possibly, our early STEEP research on the cost-ineffectiveness of the 1985 CRP county bid caps contributed, along with similar work elsewhere, to the more cost-effective signup criteria based on environmental benefits.

It is also important to adequately fund conservation programs if they are not to die after they are enacted. The landmark conservation programs of the 1985 FSA greatly increased the workload of the Natural Resource Conservation Service. However, NRCS staff and funding have not kept pace with the workload. This has caused delays and frustration for the agency and for farmers. Funding shortages have also delayed the wetland protection provisions in the 1990 and 1996 Farm Bills.

Policymakers are increasingly realizing that economics is not the only or, in some cases, the primary motivation in farmers' decisions about land use. For some of the research studies discussed in this chapter, there is substantial deviation between projected decisions based upon profit maximization and actual farmer behavior. For example, where selective enrollment of fields in CRP was projected as the most profitable means of meeting conservation compliance, farmers often have not enrolled in the program. Growers may have been reluctant to enter into land retirement contracts with the government for a 10-year period despite forecasts of short-run profitability. Growers may have desired to retain the flexibility to accommodate changes in market prices and government programs over the 10-year horizon. Desires to continue farming land for weed control, overall machinery management, and other whole farm considerations may have also been important.

Some policy analysts (Nowak, 1992) have argued that it is increasingly important to target government programs to growers in particular business situations as well as to their resource bases. Such an approach will identify the reasons growers are unable or unwilling to participate in particular conservation programs. The type of technical assistance, funding assistance, and extension programs will differ depending upon the reasons for nonparticipation.

Other analysts have stressed the need for greater state and local involvement in conservation programs. As an example, the nitrogen reduction program administered by the state of Iowa has been a major success story. Another is the groundwater protection program administered by states surrounding the Chesapeake Bay.

The diverse climates, soils, and topographies within the Pacific Northwest states produce a diverse set of agricultural enterprises and potential environmental problems. In the future, the challenge to resource agencies in the Pacific Northwest will likely be to design effective local and state policies to handle threats to the region's soil, water, and air while maintaining the economic vitality of the region's agriculture. Such a regional approach may be more effective than national policies, which tend to treat all regions and all farmers alike.

REFERENCES

Alston, J.M., and B.M. Hurd, 1990. Some neglected social costs of government spending and farm programs. *Am. J. Ag. Econ.* 72(1), 149–156.

Beus, C. et al., Eds., 1990. Prospects for Sustainable Agriculture in the Palouse. Farmer Experiences and Viewpoints, Res. Bull. No. XB1016, College of Agricultural and Home Economics, Washington State University, Pullman.

Ek, C.W., 1989. Farm Program Flexibility: An Analysis of the Triple Base Option. Congress of the U.S., Congressional Budget Office, Washington, D.C.

Ellis, J.R., J. Chivilicek, and D. Roe, 1989. Targeting highly erodible cropland for retirement: a program to supplement the conservation reserve program, paper presented at the summer meetings of the Western Agric. Econ. Assoc., Coeur D'Alene, ID, July 9–11.

Goldstein, W. A. and D. L. Young, 1987. All agronomic and economic comparison of a conventional and a low-input cropping system in the Palouse, *Am. J. Alter. Agric.* 2:51–56.

National Research Council, 1989. Alternative Agriculture. National Academy Press, Washington, D.C.

Nowak, P., 1992. Why farmers adopt production technology, *J. Soil Water Conserv.*, 47(1), 14–16.

Painter, K.M. and D.L. Young, 1993. Social welfare impacts of sustainable farming technology and agricultural policy reform, Working paper, Agricultural Economics Department, Washington State University, Pullman.

Taylor, M. L., R. M. Adams, and S. F. Miller, 1992. Farm-level response to agricultural effluent control strategies: the case of the Williamette Valley, Western J. Agric. Econ. 17(1):173–185.

U.S. General Accounting Office, 1989. Farm Programs: Conservation Reserve Program Could Be Less Costly and More Effective, GAO/RCED-90-13, Washington, D.C.

Wu, S., D.J. Walker, and M.A. Brusven, 1997. Economic and Environmental Implacts of Planting Flexibility and Conservation Compliance: Lessons From the 1985 and 1990 Farm Bills for Future Farm Legislation. *Agric. and Resource Econ. Rev.* 26(2), 216–228.

Young, D. L., A. Bechtel, and R. Coupal, 1994. Comparing Performance of the 1985 and 1990 Conservation Reserve Programs in the West. *J. Soil and Water Conservation.* 49(5), 484–487.

Young, D. L., and K. M. Painter, 1990. Farm Program Impacts on Incentives for Green Manure Rotations. *Am. J. Alter. Agric.* 5(3), 99–105.

Young, D. L., D. J. Walker, and P. L. Kanjo, 1991. Cost effectiveness and equity aspects of soil conservation programs in a highly erodible region, *Am. J. Agric. Econ.* 73(4), 1053–1062.

11 Transferring Conservation Farming Technologies to Producers

Roger J. Veseth and Donald J. Wysocki

CONTENTS

11.1 BACKGROUND AND INTRODUCTION

The increasing success and adoption of conservation tillage systems throughout the Northwest over the past 20 years is due largely to the adoption of new crop production technologies. Producers are realizing the importance of keeping up to date on new research technology to develop and maintain successful conservation tillage systems.

Much of this new technology on conservation tillage in the Northwest has been developed through STEEP and closely related research efforts. Keeping up to date on research conducted through a program such as STEEP, however, was difficult because of the size and complexity of the research effort. Over 100 scientists from the land grant universities and the USDA-Agricultural Research Service (ARS) in

0-8493-1185-3/99/$0.00+$.50

the three states have conducted as many as 50 research projects each year, involving more than 12 scientific disciplines related to crop and resource management. A focused educational effort was needed to help interpret and integrate research results into management systems for applicable agroclimatic regions, and make the technology available to Northwest producers.

During its inception in the mid-1970s, STEEP was designed as a research program. It was assumed that the existing extension system could handle the educational efforts. After several years of research in development of new conservation technologies, it became obvious that research results were not reaching producers. Although efforts were made by existing extension, NRCS, and research personnel, lack of personnel, time commitments, and resources often limited information transfer. This resulted in some long time lags between the development of components of new farming systems by researchers and the availability of that technology for application on the farm. In addition, the identity of the research developments as part of the STEEP effort was often lost in the process.

After 7 years, the STEEP program came under increasing criticism for being a "well-kept secret." Cooperative Extension in all three states was also criticized by the STEEP researchers for not doing the job of disseminating STEEP technology so it could be implemented. It was obvious that the extension component of STEEP had been overlooked in initial planning. Perhaps it was not necessary to have strong extension involvement the first few years, but after 7 years it was evident that STEEP was in trouble because of a lack of an extension component.

11.2   STEEP EXTENSION PROGRAM

11.2.1   FUNDING

In the fall of 1982, a 3-year grant from the Federal Extension Service helped initiate a tristate extension program to disseminate STEEP research information. This small, pilot project grant provided much of the salary budget for the new program. As was the case with the STEEP research program itself, federal funding for the STEEP Extension Program resulted largely from sustained efforts of Northwest wheat growers to convince their congressional delegates of the need for a STEEP extension component. The rest of the program budget was funded largely by Cooperative Extension in Idaho, Oregon, and Washington. Beginning in 1985–86 fiscal year, grants from the wheat commissions in Oregon, Washington, and Idaho provided funding for much of the operational expenses of the STEEP Extension Program. These grants allowed the continuation of STEEP Extension after expiration of the 3-year federal extension grant in 1985. Cooperative Extension in the three states has provided all the salaries and part of the operating budgets since 1987, with supplemental grants from state wheat commissions, STEEP II, STEEP III, and other sources for educational materials and programs.

11.2.2   EXTENSION POSITIONS FOCUSING ON STEEP TECHNOLOGY TRANSFER

Just as the STEEP research program was innovative in working across state lines and being an interagency, multidisciplinary effort, STEEP extension was also revolutionary. Two extension positions were set up with geographic areas of responsibility that crossed state lines and involved a complete crop management approach. Beginning as temporary positions on grant funding, they were later changed to appropriated state extension funding.

Initial efforts of the STEEP Extension Program focused on the Columbia Plateau and the Palouse and Nez Perce Prairie regions of northeastern Oregon, eastern Washington, and northern Idaho. Cropland in these regions sustains one of the highest soil erosion rates in the nation. However, a high emphasis was placed on making educational material accessible to producers in all applicable cropland areas of the Northwest.

One extension specialist position (currently held by Veseth) is located at Moscow, Idaho and is primarily responsible for dryland cropping areas of eastern Washington and northern Idaho. It is a joint position between Washington State University and University of Idaho. The other specialist position (currently held by Wysocki) is located at the Oregon State University Columbia Basin Agricultural Research Center near Pendleton. This facility, which is shared with the USDA-ARS Columbia Plateau Conservation Research Center, was composed only of researchers up until 1982. The primary geographic area of responsibility for the position covers north-central and northeastern Oregon and two adjacent counties in southeast Washington.

The STEEP extension effort expanded during the STEEP II and STEEP III programs. A Pacific Northwest (PNW) STEEP Cropping Systems Specialist Team was established with five other agronomy and soils specialists in the three states. The team has collaborated on several publications, a STEEP II on-farm testing project, the establishment of a STEEP World Wide Web Home Page, and other educational materials and programs in the region.

11.2.3   COOPERATIVE TEAM APPROACH

STEEP extension has been a cooperative-education effort. Existing producer information networks of county agricultural extension agents, conservation districts, USDA-NRCS personnel, grain-grower associations, crop improvement associations, and agricultural service industries have been utilized in dissemination of the new conservation farming technology. These networks provide pathways for dissemination of information through their local grower newsletters and other publications, and conservation farming/crop production meetings, workshops, and field tours. Cooperative efforts between STEEP extension and these agencies and groups have been critical for effective information transfer to producers.

11.3   METHODS OF TECHNOLOGY TRANSFER

A wide variety of media have been utilized for technology transfer, including newsletters, articles for farm magazines and newspapers, extension and popular publications, audiovisuals, meeting presentations, conferences and workshops, on-farm testing and field demonstrations, and educational exhibits. Educational materials were developed by the extension specialists who focused on the STEEP program, in cooperation with STEEP researchers, other extension specialists, county agricultural agents, and personnel from related agencies and the agricultural industry.

11.3.1 NEWSLETTERS

Newsletters have been a valuable and effective part of STEEP extension. The initial STEEP extension newsletters, developed separately by the two extension specialists for their respective regions, evolved into the joint *PNW STEEP Extension Conservation Farming Update* in 1986.

The primary goal of the *Update* and related educational efforts has been to accelerate Northwest producer access to and adaptation of new technologies. New technology has been addressed from three perspectives: (1) what agroclimatic areas it applies to; (2) how producers might incorporate it into their farming systems; and (3) how it relates to other management considerations. A secondary role of the *Update* has been to increase STEEP program visibility and public awareness of its importance as a technology source for effective, profitable conservation farming systems in the region.

The two specialists worked closely with STEEP researchers and other extension specialists in preparing articles and other publications on research findings for specific management topics. The mailing list has grown from a few hundred in the early years to over 2700 as an increasing number of growers and agricultural industry personnel asked to be on the mailing list. Many of the articles have been condensed or reprinted entirely in local newsletters by county agents, conservation districts, agricultural magazines and newspapers, and other grower information networks. This broader technology transfer network has been critical for increasing producer awareness and adaptation of STEEP technology.

11.3.2 FARM MAGAZINE AND NEWSPAPER ARTICLES

A STEEP survey of growers showed that local and regional farm magazines and newspapers are an important source of information. Thus, farm magazines and newspapers have provided an opportunity to reach a much wider audience than was possible through newsletters and other publications. From 1983 through 1998, an average of 28 STEEP extension articles on a wide range of topics related to conservation farming systems were carried each year in eight regional farm magazines and newspapers in the Northwest with a combined circulation of over 200,000.

11.3.3 EXTENSION PUBLICATIONS

Extension publications from individual states and PNW extension publications have been used to compile, integrate, and disseminate new technologies from STEEP and related research programs. These publications give in-depth reviews of new research developments and state-of-the-art management strategies. They provide a permanent reference source for STEEP program information and are available from university extension publication offices and from county extension offices in the three states.

An article appeared in a national extension publication (Reinertsen et al., 1983d) on the role that the STEEP Extension Program plays in the overall STEEP effort.

This helped to enhance national recognition of the STEEP program, which has become a national model for cooperative efforts to solve regional environmental and economic problems.

A PNW extension publication highlighted research on "slot-mulching," as an innovative method to prevent runoff, increase water storage, and facilitate the use of no-till systems (Reinertsen et al., 1983c). This technique involves collecting and compacting combine straw residue into shallow trenches on the field contour. Publications on the effect of tillage systems on yields of spring barley (Reinertsen et al., 1983a) and spring wheat (Reinertsen et al., 1983b) were published by both Washington State University and the University of Idaho.

Eight extension publications were published by Oregon State University in a STEEP Special Rpt. (SR) series. These include topics on residue management (Maxwell and Ramig, 1984; Maxwell, 1984), tillage systems (Maxwell et al., 1984a, b), chemical fallow (Rydrych and Maxwell, 1985), cropping on shallow soils (Vomocil and Ramig, 1984), soil freezing (Vomocil et al., 1984) and erosion in western Oregon (Istok and Vomocil, 1985a, b).

A series of PNW extension bulletins on crop management in conservation tillage systems was initiated in 1986. Topics have included fertilizer placement for cereal root access in conservation tillage (Veseth et al., 1986b), minimum tillage and no-till as effective conservation farming systems (Veseth et al., 1986c), and combine residue distribution for successful conservation tillage systems (Veseth et al., 1986a). A PNW extension publication was developed on agronomic zones for the dryland PNW (Douglas et al., 1990) to assist in determining the areas of application for new conservation farming technologies.

A *Pacific Northwest Conservation Tillage Handbook* (Veseth and Wysocki, 1989) was published in 1989. The purpose of the *Handbook* was to help consolidate an extensive amount of new STEEP and related management technology for conservation tillage systems into an organized resource guide to increase the accessibility of the information. The *Handbook* is a large, three-ring binder to permit continued additions and revisions so it can be maintained as a current, up-to-date reference in the future.

The *Handbook* consists primarily of *PNW Conservation Tillage Handbook Series* publications and other PNW extension bulletins. All have been published or revised since 1984. They are in-depth publications developed from the perspective of how the new research technologies can fit into producer management systems for specific agronomic areas.

The *Handbook* was initially released with 98 *Handbook Series* publications. There were 32 new *Handbook Series* publications added to the *Handbook* by 1996 (Veseth, 1989a through 1992b; Wysocki, 1989 through 1990c; Wysocki and Veseth, 1991; Veseth et al., 1992; 1993; 1994; Pike et al., 1993; Schillinger et al., 1995; Young et al., 1995; Veseth et al., 1996). The *Pacific Northwest STEEP Conservation Farming Update* newsletter serves as a mechanism for distributing new *Handbook Series* publications.

Everyone who purchases the *Handbook* (and returns the enclosed updating card) is added to the *Update* mailing list. *Handbook Series* publications are three-hole, punched, and ready for insertion at the beginning of the ten specified chapters in the *Handbook*. Updated tables of content are provided periodically to include new additions. The *Handbook* and new *Handbook Series* are also available on the Internet (http://pnwsteep.wsu.edu).

11.3.4   POPULAR PUBLICATION

Wheat Health Management (Cook and Veseth, 1991) was the first book in a new Plant Health Management Series by the American Phytopathological Society (APS) Press. This comprehensive book integrates all important facets of wheat health management into a decision framework to help growers develop more efficient, environmentally sound production systems that optimize yields within the constraints of the environment. It was written to help guide wheat health managers—producers, fieldmen, farm advisors, and extension and other agricultural service and support personnel—to an understanding of the basic concepts and approaches to wheat health management. The unique "holistic" approach of this book focuses on the whole cropping system—not just on the wheat plant or on individual management choices apart from interactions within the overall cropping system.

Although *Wheat Health Management* is a book of North American scope, it incorporates much of the new STEEP technology for improving wheat health and production potential in the PNW. STEEP researchers have been among the national leaders in developing wheat health management technologies for conservation cropping systems.

There is a growing awareness in this country and worldwide that agriculture must become more productive and efficient yet also be sustainable and ecologically sound in the long term. To accomplish these sometimes seemingly contradictory goals, agriculture must become increasingly more sophisticated and technical. *Wheat Health Management* is intended to help producers develop comprehensive cropping system strategies that focus on both profitability and environmental protection. For more information on the book, call APS Press toll-free at 1-800-328-7560.

11.3.5   WORLD WIDE WEB HOME PAGE

The World Wide Web (WWW) is becoming an increasingly important source of technologies on farming systems. A WWW STEEP home page on dryland farming systems technology for improved profitability and erosion control in the Northwest was initiated in the fall of 1995 by the PNW STEEP Cropping Systems Specialist Team. It is titled "PNW STEEP Conservation Farming Systems Information Source" (http://pnwsteep.wsu.edu). All of the PNW Conservation Tillage Handbook Series are being put on the home page. Other features include new issues of the *PNW STEEP Extension Conservation Farming Update* newsletters, conference proceedings, coming events, and links to other sites.

11.3.6   AUDIOVISUALS

Audiovisual materials can be an effective method of presenting information and can be made available to more audiences without the time and travel expense of appearing in person. Several slide/cassette tape and video series have been prepared. With funding from the original Federal Extension grant, a slide/cassette tape series was developed on conservation tillage (Engle, 1983). The series was initially used in conjunction with a teleconference between STEEP researchers and extension specialists at Pullman, and producers and county agents at seven eastern Washington locations. A slide series on patterns of wheat plant development (Maxwell et al., 1983) was developed in 1983 with ARS STEEP researchers at Pendleton, Oregon.

Two slide/cassette tape series and videotapes (of the slide/tape series) were completed in 1986. These summarized PNW extension publications PNW 283 on fertilizer placement (Veseth, 1986a) and PNW 275 on effective conservation tillage systems (Veseth, 1986b).

Two videos highlight management strategies for improving the success of conservation tillage systems. One on profitable conservation cropping systems (Boerboom and Veseth, 1991) provides an overview of a 6-year project associated with the STEEP program on integrated pest management for conservation cropping systems. The other addresses managing the green bridge for root disease control in conservation tillage (Veseth and Cook, 1993).

11.3.7   MEETING PRESENTATIONS

Growers and others seeking knowledge on new crop management technologies for conservation tillage are attracted to meetings, workshops, and conferences around the Northwest. Thus, meeting presentations are a good way to disseminate information from STEEP research. These meetings commonly include county extension crop management workshops and local, regional, and state meetings of grain producers, conservation districts, and agricultural support industries. Each year since 1983, the two extension specialists focusing on STEEP extension made presentations to an average of over 1580 producers at about 25 meetings. Presentation topics have covered research-based information for developing effective conservation farming systems.

11.3.8   ORGANIZING CONFERENCES AND WORKSHOPS

Regional conservation tillage conferences and workshops have provided excellent forums for researchers, producers, and industry to share conservation tillage information. Organizing these events has been a worthwhile educational endeavor. Ten Inland Northwest Conservation Tillage Conferences have been held in Moscow, Idaho and Pullman, Washington. Four similar Conservation Tillage Workshops were conducted in Pendleton or the Dalles, Oregon. Attendance at the conferences averaged about 300, ranging from 100 up to 450. Many of the conferences also included extensive exhibitions of conservation tillage equipment and products. Most also included detailed proceedings of the conference presentations.

Beginning in 1993, STEEP extension specialists helped change the Annual Reviews of PNW STEEP Research into PNW Conservation Farming Conferences with expanded publicity and program agenda designed for a grower audience.

Conference attendance grew from the typical 100-200 to 300-400. Another major change in conferences was initiated by the STEEP Extension Team in 1998 with the first Northwest Direct Seed Cropping Systems Conference. New features included co-sponsorship and participation by agricultural industry, and the addition of international speakers and a conference proceedings. The attendance of nearly 900 growers and high evaluation ratings indicated that the new type of conference was well received and that there is increasing interest in these new cropping systems.

Annual reviews of conservation farming research by USDA-ARS and Oregon State University scientists in northeast Oregon have been conducted at Pendleton, Oregon in most years since 1985. The target audience has been primarily county extension agents and NRCS staff in Oregon and Washington, and the purpose has been to assist them in advising growers and the agricultural support industry on conservation farming systems.

11.3.9   ON-FARM TESTING AND FIELD DEMONSTRATIONS

On-farm tests and field demonstrations give producers firsthand observations of conservation tillage equipment and systems. They provide an opportunity to incorporate different management components from STEEP research into production systems on a field-scale level with grower equipment.

From 1983 through 1990, more than 14 field demonstration trials were conducted in the inland Northwest area of the three states. Examples of topics included no-till spring barley and spring wheat, Paraplow tillage, slot-mulching, annual cropping with nine different crop rotations and alternative crops in traditional cropfallow regions, uphill plowing, subsoil tillage practices. Tours were conducted for area farmers, extension agents, conservation districts, and NRCS and industry personnel.

A seeding demonstration of 12 no-till and minimum-tillage drills attracted over 350 people to a 1985 field day near Pullman. A residue management demonstration was conducted for about 250 producers during the 1988 annual field day of the Columbia Basin Agricultural Research Center at Pendleton, Oregon. Through various tillage operations, 20 plots were prepared to show surface residue cover from 12 to 95%.

When STEEP II began in 1991, efforts on on-farm testing were greatly expanded. The focus of a 5-year STEEP II on-farm testing project was on helping producers and agricultural support personnel evaluate new conservation farming practices with scientifically sound on-farm test designs using producers' field equipment. On-farm test methodologies were developed for PNW field landscape conditions (Wuest et al., 1994a). An on-farm test record form was developed to aid in data collection (Wuest et al., 1995a). A publication was also completed on using on-farm testing for variety selection (Wuest et al., 1995b). The number of on-farm tests reported in annual *PNW On-Farm Test Results* publications increased from 25 in 1992 (Wuest et al., 1992) to 43 in 1993 (Wuest et al., 1993), 65 in 1994 (Wuest et al., 1994b), and 51 in 1995 (McClellan et al., 1996). Since the on-farm testing project began, the use of replicated on-farm tests with field-scale equipment has become a more accepted or "standard" approach to field evaluation of new farming technologies by university and industry researchers, agricultural support personnel and producers.

11.3.10 Educational Resource Exhibits

Displays of information resources have been used to publicize the STEEP research and extension program and improve producer access to new conservation farming technology. The first display was prepared at the request of the U.S. Congressional Agriculture Committee for a National Agricultural Research Fair held in Washington, D.C. in 1983.

A new display was developed for a 1998 Washington D.C. exhibition titled "How Agricultural Science Research Serves the Nation" sponsored by the National Assoc. of State Universities and Land-Grant Colleges (NASULGC). The displays has been expanded as a resource exhibits to highlight STEEP extension newsletters, publications, and other educational resources on conservation farming systems. The resource exhibit has been continually updated and viewed by an average audience of 1800 at about seven meetings and field days each year since 1984.

11.4 SUMMARY

STEEP extension has helped to speed the transfer of new STEEP technologies to Northwest producers. The need for extension programs in conservation farming has increased as STEEP II, STEEP III, and related research programs develop new technologies for management of conservation tillage systems in the Northwest. This need is most critical for the ultimate user, the producer, but is also vital for STEEP researchers, other extension specialists and agents, Agricultural service industry personnel, conservation districts, NRCS staff, and personnel from other agricultural agencies and groups who participate in developing new equipment, products, and management practices for conservation farming systems, or advise growers in utilizing these technologies. The need for a focused research and extension efforts on conservation farming systems will grow as producers are increasingly faced with environmental and economic challenges in the future.

REFERENCES

Boerboom, C. and R.J. Veseth, 1991. Profitable conservation cropping systems: insights from the 6-year USDA-ARS IPM-conservation cropping systems project, Washington State University Cooperative Extension video VT0029, Pullman.

Cook, R.J. and R.J. Veseth, 1991. *Wheat Health Management*, American Phytopathological Society Press, St. Paul, Minnesota.

Douglas, L.C., D.J. Wysocki, J.F. Zuzel, R.W. Rickman, and B.L. Klepper, 1990. Agronomic zones for the dryland Pacific Northwest, *PNW Ext. Bull.*, PNW 354.

Engle, C., 1983. Conservation Tillage in the Inland Northwest, Washington State University Cooperative Extension unnumbered slide/cassette series, Pullman.

Istok, J. and J. A. Vomocil, 1985a. Drainage for Erosion Control in Western Oregon, Oregon State University Extension Service, STEEP Spec. Rpt. SR 733.

Istok, J. and J. A. Vomocil, 1985b. Some Willamette Hillslope Erosion Rates, Oregon State University Extension Service, STEEP Spec. Rpt. SR 736.

Maxwell, D.C., E.L. Klepper, and R.W. Rickman, 1983. Patterns of Wheat Plant Development, Oregon State University Extension Service (Unnumbered slide series).

Maxwell, D.C., 1984. Improving Cereal Residue Distribution for No-Till, Oregon State University Extension Service, STEEP Spec. Rpt. SR 719.

Maxwell, D.C. and R.E. Ramig, 1984. Small Grain Residue Management, Oregon State University Extension Service, STEEP Spec. Rpt. SR 699.

Maxwell, D.C., R. Costa, G. Brog, D. Dickens, and S. Siles, 1984a. No-Till Management Systems vs. Conventional Tillage Systems: A Cost Comparison, Oregon State University Extension Service (an unnumbered enterprise cost study).

Maxwell, D.C., F.V. Pumphrey, and D.C. Hane, 1984b. Conservation Tillage for Field Corn Production, Oregon State University Extension Service, STEEP Spec. Rpt. SR 710.

McClellan, R., B.C. Miller, T. Fiez, R.J. Veseth, S.O. Guy, D.J. Wysocki, and R.S. Karow, 1996. 1995 Pacific Northwest On-Farm Test Results from the Idaho, Oregon and Washington STEEP II On-Farm Testing Project, Washington State University Cooperative Extension Crop and Soil Sciences Department, Technical Rpt. 96-1, Pullman.

Pike, K., R.J. Veseth, B.C. Miller, R. Schirman, L. Smith, and H. Homan, 1993. Hessian fly management in conservation tillage systems for the Inland Pacific Northwest, *PNW Extension Conservation Tillage Handbook Series*, Chap. 8, No. 15.

Reinertsen, S.A., A.J. Ciha, and C.F. Engle, 1983a. Yield of four spring barley varieties under conventional, minimum, and no-tillage systems, *Washington State Univ. Ext. Bull.*, 1094, University of Idaho Current Information Series 687.

Reinertsen, S.A., A.J. Ciha, and C.F. Engle, 1983b. Yield of four spring wheat varieties under conventional, minimum, and no-tillage systems, *Washington State Univ. Ext. Bull.*, 1094. University of Idaho Current Information Series 689.

Reinertsen, S.A., K.E. Saxton, and C.F. Engle, 1983c. Slot mulching for residue management and runoff control. 1983, *PNW Ext. Bull.*, 231.

Reinertsen, S.A., R.E. McDole, C.F. Engle, and J.A. Vomocil. 1983d. Extension's role in STEEP, *USDA Ext. Rev.*, Spring, 20–21.

Rydrych, D.J. and D.C. Maxwell, 1985. Chemical Fallow in Oregon Dryland Grain, Oregon State University Extension Service, STEEP Spec. Rpt. SR 746.

Schillinger, W., D. Wilkins, and R. Veseth, 1995. Deep rippping fall-planted wheat after fallow to improve infiltration and reduce erosion, *PNW Conservation Tillage Handbook Series*, Chap. 2, No. 16, PNW extension publication in Idaho, Oregon, and Washington.

Veseth, R.J., 1986a. Fertilizer band location for cereal root access in no-till and minimum tillage systems, University of Idaho Cooperative Extension System video 615.

Veseth, R.J., 1986b. Minimum tillage and no till: effective conservation farming systems, University of Idaho Cooperative Extension System video 616.

Veseth, R.J., 1989a. Soil sampling in fertilizer-banded fields, *PNW Extension Conservation Tillage Handbook Series*, Chap. 6, No. 14.

Veseth, R.J., 1989b. Erosion makes soils more erodible, *PNW Extension Conservation Tillage Handbook Series*, Chap. 1, No. 10.

Veseth, R.J., 1989c. Surface residue reduces overwinter evaporation, *PNW Extension Conservation Tillage Handbook Series*, Chap. 3, No. 14.

Veseth, R.J., 1990a. Yield mapping could improve crop and resource management, *PNW Extension Conservation Tillage Handbook Series*, Chap. 10, No. 7.

Veseth, R.J., 1990b. Winter wheat nitrogen management in the 18- to 25-inch precipitation zone, *PNW Extension Conservation Tillage Handbook Series*, Chap. 6, No. 15.

Veseth, R.J., 1990c. Winter lentil could provide conservation tillage option, *PNW Extension Conservation Tillage Handbook Series*, Chap. 8, No. 12.

Veseth, R.J., 1990d. Fertilizer placement-row spacing effects on wild oat, *PNW Extension Conservation Tillage Handbook Series*, Chap. 5, No. 14.

Veseth, R.J., 1990e. Radioactive fallout provides estimator for soil erosion, *PNW Extension Conservation Tillage Handbook Series*, Chap. 1, No. 11.

Veseth, R.J., 1990f. Winter rapeseed recropping consideration, *PNW Extension Conservation Tillage Handbook Series*, Chap. 8, No. 14.

Veseth, R.J., 1990g. No-till winter wheat after green manure legumes, *PNW Extension Conservation Tillage Handbook Series*, Chap. 2, No. 15.

Veseth, R.J., 1991a. Fertilizer placement can reduce root disease effects, *PNW Extension Conservation Tillage Handbook Series*, Chap. 4, No. 15.

Veseth, R.J., 1991b. Fertilizer placement can reduce root disease effects, *PNW Extension Conservation Tillage Handbook Series*, Chap. 4, No. 15.

Veseth, R.J., 1992a. Green bridge key to root disease, *PNW Extension Conservation Tillage Handbook Series*, Chap. 4, No. 16.

Veseth, R.J., 1992b. Green bridge control starts in the fall, *PNW Extension Conservation Tillage Handbook Series*, Chap. 4, No. 18.

Veseth, R.J. and R.J. Cook, 1993. Managing the green bridge: root disease control in conservation tillage, Washington State University Cooperative Extension video VT0040.

Veseth R.J. and D.J. Wysocki, 1989. *PNW Extension Conservation Tillage Handbook*, Pacific Northwest Extension Publication in Idaho, Oregon, and Washington, Moscow, ID.

Veseth, R.J., C. Engle, J. Vomocil, and R. McDole, 1986a. Uniform combine residue distribution for successful no-till and minimum tillage systems, *PNW Ext. Bull.* 297 and *PNW Extension Conservation Tillage Handbook Series*, Chap. 2, No. 4.

Veseth, R., R. McDole, C. Engle, and J. Vomocil, 1986b, Fertilizer band location for cereal root access in no-till and minimum tillage systems, *PNW Ext. Bull.*, 283 and *PNW Extension Conservation Tillage Handbook Series*, Chap. 3, No. 7.

Veseth, R.J., J. Vomocil, R. McDole, and C. Engle. 1986c. Minimum tillage and no till: effective conservation farming systems, *PNW Ext. Bull.* 275 and *PNW Extension Conservation Tillage Handbook Series*, Chap. 6, No. 4.

Veseth, R.J., K. Saxton, and D. McCool, 1992. Tillage and residue management strategies for variable cropland, *PNW Extension Conservation Tillage Handbook Series*, Chap. 3, No. 18.

Veseth, R.J., B.C. Miller, S.O. Guy, D.J. Wysocki, T. Murray, R. Smiley, and M. Wiese, 1993. Managing Cephalosporium stripe in conservation tillage systems, *PNW Conservation Tillage Handbook Series*, Chap. 4, No. 17, Sept.

Veseth, R.J., A. Ogg, D. Thill, D. Ball, D. Wysocki, F. Bailey, T. Gohlke, and H. Riehle, 1994. Managing downy brome under conservation tillage systems in the inland Northwest crop-fallow region, *PNW Extension Conservation Tillage Handbook Series*, Chap. 5, No. 15.

Veseth, R.J., B. Miller, T. Fiez, T. Walters, and H. Schafer, 1996. Returning CRP land to crop production—a summary of 1994–1996 research trials in Washington State, *PNW Conservation Tillage Handbook Series*, Chap. 2, No. 6, PNW Extension Publication in Idaho, Oregon, and Washington.

Vomocil, J. A. and R. Ramig, 1984. Annual Cropping on Shallow Soils, Oregon State University Extension Service, STEEP Spec. Rpt. SR 701.

Vomocil, J. A., J. Zuzel, J. Pikul, and D. Baldwin, 1984. Stubble Management Influences Soil Freezing, Oregon State University Extension Service, STEEP Spec. Rpt. SR 700.

Wuest, S.B., B.C. Miller, R.J. Veseth, S.O. Guy, D.J. Wysocki, and R.S. Karow, 1992. 1992 Pacific Northwest On-Farm Test Results, Department of Crop and Soil Sciences, Technical Rpt. 95-1, Washington State University, Pullman.

Wuest, S.B., V.C. Miller, J.R. Alldredge, S.O. Guy, R.S. Karow, R.J. Veseth, and D.J. Wysocki, 1994a. Increasing Plot Length Reduces Experimental Error of On-Farm Tests, 1992 Pacific Northwest On-Farm Test Results, Department of Crop and Soil Sciences, Technical Rpt. 95-1, Washington State University, Pullman.

Wuest, S.B., B.C. Miller, R.J. Veseth, S.O. Guy, D.J. Wysocki, and R.S. Karow, 1993. 1993 Pacific Northwest On-Farm Test Results, Department of Crop and Soil Sciences, Technical Rpt. 95-1, Washington State University, Pullman.

Wuest, S.B., B.C. Miller, R.J. Veseth, S.O. Guy, D.J. Wysocki, and R.S. Karow, 1994b. 1994 Pacific Northwest On-Farm Test Results, Department of Crop and Soil Sciences, Technical Rpt. 95-1, Washington State University, Pullman.

Wuest, S.B., B.C. Miller, R.J. Veseth, S.O. Guy, D.J. Wysocki, and R.S. Karow, 1995a. On-Farm Test Record Form, *PNW Ext. Bull.* 487, Fact Sheet 1.

Wuest, S.B., B.C. Miller, R.S. Karow, S.O. Guy, R.J. Veseth, and D.J. Wysocki, 1995b. Using On-Farm Testing for Variety Selection, *PNW Ext. Bull.*, PNW 487, Fact Sheet 2.

Wysocki, D.J., 1989. Runoff and erosion events in the inland Northwest, *PNW Extension Conservation Tillage Handbook Series*, Chap. 3, No. 15.

Wysocki, D.J., 1990a. Conservation farming and sustainability, *PNW Extension Conservation Tillage Handbook Series*, Chap. 1, No. 12.

Wysocki, D.J., 1990b. Developing resistance to Russian wheat aphid, *PNW Extension Conservation Tillage Handbook Series*, Chap. 8, No. 13.

Wysocki, D.J., 1990c. Wheat residue composition, decomposition and management, *PNW Extension Conservation Tillage Handbook Series*, Chap. 3, No. 16.

Wysocki, D.J. and R.J. Veseth, 1991. Tillage and stubble management for water conservation, *PNW Extension Conservation Tillage Handbook Series*, Chap. 5, No. 16.

Young, F.L., R.J. Veseth, D. Thill, W. Schillinger, and D. Ball, 1995. Russian thistle management under conservation systems in Pacific Northwest crop-fallow regions, *PNW Extension Conservation Tillage Handbook Series*, Chap. 5, No.16 and *PNW Extension Bull.* 492.

Index